Proceedings

Ein stetig steigender Fundus an Informationen ist heute notwendig, um die immer komplexer werdende Technik heutiger Kraftfahrzeuge zu verstehen. Funktionen, Arbeitsweise, Komponenten und Systeme entwickeln sich rasant. In immer schnelleren Zyklen verbreitet sich aktuelles Wissen gerade aus Konferenzen, Tagungen und Symposien in die Fachwelt. Den raschen Zugriff auf diese Informationen bietet diese Reihe Proceedings, die sich zur Aufgabe gestellt hat, das zum Verständnis topaktueller Technik rund um das Automobil erforderliche spezielle Wissen in der Systematik aus Konferenzen und Tagungen zusammen zu stellen und als Buch in Springer.com wie auch elektronisch in Springer Link und Springer Professional bereit zu stellen. Die Reihe wendet sich an Fahrzeug- und Motoreningenieure sowie Studierende, die aktuelles Fachwissen im Zusammenhang mit Fragestellungen ihres Arbeitsfeldes suchen. Professoren und Dozenten an Universitäten und Hochschulen mit Schwerpunkt Kraftfahrzeug- und Motorentechnik finden hier die Zusammenstellung von Veranstaltungen, die sie selber nicht besuchen konnten. Gutachtern, Forschern und Entwicklungsingenieuren in der Automobil- und Zulieferindustrie sowie Dienstleistern können die Proceedings wertvolle Antworten auf topaktuelle Fragen geben.

Today, a steadily growing store of information is called for in order to understand the increasingly complex technologies used in modern automobiles. Functions, modes of operation, components and systems are rapidly evolving, while at the same time the latest expertise is disseminated directly from conferences, congresses and symposia to the professional world in ever-faster cycles. This series of proceedings offers rapid access to this information, gathering the specific knowledge needed to keep up with cutting-edge advances in automotive technologies, employing the same systematic approach used at conferences and congresses and presenting it in print (available at Springer.com) and electronic (at Springer Link and Springer Professional) formats. The series addresses the needs of automotive engineers, motor design engineers and students looking for the latest expertise in connection with key questions in their field, while professors and instructors working in the areas of automotive and motor design engineering will also find summaries of industry events they weren't able to attend. The proceedings also offer valuable answers to the topical questions that concern assessors, researchers and developmental engineers in the automotive and supplier industry, as well as service providers.

Weitere Bände in der Reihe http://www.springer.com/series/13360

Wolfgang Siebenpfeiffer
(Hrsg.)

Automotive Acoustics Conference 2019

5. Internationale ATZ-Fachtagung
Fahrzeugakustik

Springer Vieweg

Hrsg.
Wolfgang Siebenpfeiffer
Stuttgart, Deutschland

ISSN 2198-7432 ISSN 2198-7440 (electronic)
Proceedings
ISBN 978-3-658-27668-3 ISBN 978-3-658-27669-0 (eBook)
https://doi.org/10.1007/978-3-658-27669-0

Die Deutsche Nationalbibliothek verzeichnet diese Publikation in der Deutschen Nationalbibliografie; detaillierte bibliografische Daten sind im Internet über http://dnb.d-nb.de abrufbar.

Verantwortlich im Verlag: Markus Braun

Springer Vieweg ist ein Imprint der eingetragenen Gesellschaft Springer Fachmedien Wiesbaden GmbH und ist ein Teil von Springer Nature.
Die Anschrift der Gesellschaft ist: Abraham-Lincoln-Str. 46, 65189 Wiesbaden, Germany

Editorial

The increased variety and complexity of today's powertrains further increases the effort required for acoustic development. As a result, acoustic experts are faced with the challenge of effectively reducing the noise of the powertrain and its components. Lightweight construction also increases the complexity of striving for a comfortable NVH level. Better measurement analyses and faster calculation methods are important tools for this and central topics in the experts' discussion. Starting with keynote lectures by leading industry executives, the 5th Automotive Acoustics Conference will cover its customary broad spectrum of themes, giving participants a comprehensive, bi-annual snapshot of the state of the art in acoustics and NVH.

In addition to 20 presentations, the 2019 program also features two parallel expert discussions aiming at the latest trends in the automotive industry: "The future of aeroacoustics in the automotive industry" and "Tire noise: an old problem that requires new methods, new products, new ideas".

The next edition of this leading global conference in the field of automotive acoustics will again take place at the renowned Swiss thinktank Gottlieb Duttweiler Institute (GDI) in Zurich/Rueschlikon under the patronage of Professor Dr. Paolo Ermanni of ETH. We look forward to meeting you in Zurich on 9 and 10 July 2019!

On behalf of the Scientific Advisory Board

Dr. Davide Caprioli
Head of Product Acoustic and Thermal Performance
Autoneum Management Ltd

Dr. Alexander Heintzel
Editor-in-Chief
ATZ | MTZ Group
Springer Nature

Inhaltsverzeichnis

NVH Refinement Issues for BEV

Léon Gavric(✉)

VP & Senior NVH Expert, Department of Quality and Engineering, PSA Groupe, Paris, France
leon.gavric@mpsa.com

Abstract. Regulatory pressures on internal combustion engines (ICEs), combined with technological improvements in electric powertrains and batteries lead to an increase of demand for electric vehicles (EVs). Both, the traditional carmakers as well as new entrants without ICE legacies are developing and putting on the market new EV models.

Worldwide sales of pure battery EVs (BEV-excluding hybrids) grew by approximately 45% in 2016. Since, BEVs become mass-market product, it is necessary that automotive acoustic engineers apprehend and evaluate related technology trends and their influences on vehicle NVH. The BEV is a radically different vehicle and it needs an updated NVH approach, which targets all relevant NVH comfort issues in a specific electric environment. As demand rises, BEV technology and design will continue to evolve, and strategic challenges will follow. Established OEMs and their traditional suppliers will need to rethink their approaches to preserve their revenue and profitability.

The ZE driving implies "silence". Consequently, some NVH performances are to be increased for BEVs. The lack of IC engine noise masking exacerbates: the noises of different systems & devices, wind noise, rolling noise and tire hiss, etc. the corresponding NVH requirements are to be strengthened, comparing to the IC engine vehicles. The completely new, but crucial NVH challenge for BEV is to control e-Powertrain whining. Disturbing noises like squeak rattle and buzz, particularly in low noise BEV context are prohibited, too.

Keywords: Battery Electric Vehicles · NVH performances · NVH design

1 Actual Market and BEV

1.1 Why do EOMs Invest Massively in BEV

Environmental, friendly car becomes societal objective due to: the ecological conscience of customer, the tax policy in favor of low consumption cars and CO_2 reduction objectives imposed by EU legislation.

The 'ecological conscience' of consumer, the CO_2 reduction imposed by legislation and the low consumption tax policy adopted by majority of countries, motivate car manufacturers to develop new generation of low CO_2 emission cars.

W. Siebenpfeiffer (Hrsg.): *Automotive Acoustics Conference 2019*, Proceedings, S. 1–9, 2020. https://doi.org/10.1007/978-3-658-27669-0_1

Concerning the CAFE The reduction of CO_2 emission is required by new European legislation. The objectives are fixed by CAFE index: "Corporate Average Fuel Economy" and expressed as quantity of CO_2 generated by average car in grams per kilometer. The averaging is performed on total car ensemble sold in all European countries by a manufacturer. The non-respect of the objectives fixed by EU will lead to financial penalties proportional to the observed deviation. By 2021, phased in from 2020, the fleet average to be achieved by all new cars is 95 g of CO_2 per kilometer (4.0 L/km). From 2019, the penalty will be €95 form the first gram of exceedance. The low emission BEVs will count double when compute the average. Moreover the percentage of low emission vehicles (BEV and PHEV) is imposed for the years to come (2025 xx% and 2030 XX%). Therefore, BEV technology seem to be unavoidable in order to reach those objectives.

CO_2 emissions from new cars should be cut by 45% by 2030 and market uptake of zero – and low-emission vehicles should accelerate, said committee MEPs on Monday.

Environment Committee MEPs proposed setting a higher target for reducing EU fleet-wide emissions for new cars by 2030: 45% (compared to the EU Commission's 30%) with an intermediate target of 20% by 2025. The draft legislation also sets similar targets for new vans.

Manufacturers whose average CO_2 emissions exceed these targets will pay a premium to the EU budget, to be used, inter alia, for up-skilling workers affected by changes in the automotive sector, MEPs agreed.

Carmakers will also have to ensure that zero- and low- emission vehicles – ZLEVs – (which emit less than 50 g CO_2/km) have a 40% market share of sales of new cars and vans by 2030, and 20% by 2025.

1.2 BEV Technology

The principle of BEV is simple: an electric motor powered by a battery replaces the Internal Combustion Engine Vehicle (ICEV) and the tank, and the vehicle is plugged to a charging spot when the battery needs to be charged (Fig. 1).

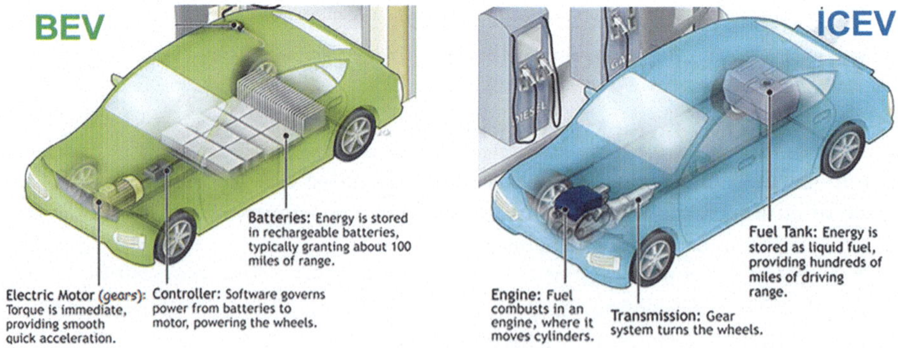

Fig. 1. BEV (Battery Electric Vehicle) vs ICEV (Internal Combustion Engine Vehicle): technology and principal devices needed for generation of motion and energy storage.

They have many advantages: they are highly efficient, do not produce tailpipe emissions which is beneficial for local air quality, they have good acceleration, can be charged overnight on low cost electricity produced by any type of power station, including renewables. However, despite these advantages, BEVs, also, face significant challenges. Electricity storage is still expensive and the charging of the battery is time consuming; this is why the range of these vehicles is limited (Fig. 2).

Fig. 2. E-powertrain and battery pack are the principal subsystems needed for electric motion.

E-powertrin comparing to IC-powertrain
It is generally of lower mass and moment of inertia, which implies lower excitation by road induced vibrations. It has a high torque, which can be observed even at low motor speeds. This may excite some driveline modes, like "tip-in / let-off" or sway mode. E-motors have a low torque irregularity. In electric drives, it is usually called "torque ripple" and generally does not exceed few percent of overall torque (~5%).

E-motors for vehicle driveline have high rotational speeds (up to 15 000 rpm). Still the inertial forces are low since there is no linear moving of masses (like pistons, valves, etc.) and consequently no unbalanced inertial forces. E-powertrain does not have idle speed vibrations neither shake due to stop & start.

Battery pack capacity and platform issues
A specific or "native" platform is needed for a high-performances and driving range of BEV. A clear gap is observed in driving range and interior space between models with native BEV platforms and those compatible with IC powertrain. Non-native or multi-energy BEV platforms force the battery into the awkward footprint of the existing IC powertrain vehicles. The battery pack in native BEV platforms, by contrast, can take a simple, rectangular shape, giving native BEVs up to twice the range—over 300 km per charge and up to approximately 400 km for the best performers, for today's battery technologies. In addition, native BEVs achieve a larger interior space (up to 10% by regression line) for the same wheelbase compared with not only non-native counterparts, but also standard ICE vehicles in the same segment.

1.3 Why Do Peoples Like BEV (… When They Like It !)

BEV are really fun and easy to drive. Nothing in the car needs to gear up so when you step on the accelerator it just goes, and it accelerates fast and from any initial speed. Conversely, you can use your brakes a lot less while driving an electric car. Moreover the breaking recaptures power to feed back to the battery. Not only does this let you go farther without recharging but it's very handy because all you have to do is lift your foot of the "gas" pedal and the car will start slowing down by itself.

Electric cars are quiet. That can make them even dangerous around pedestrians. To combat this, the legislation imposes AVAS (Alerting Vehicle Acoustic System), a devise producing alerting sound at speeds below 20 mph. Even so, BEV is infinitely quieter than a ICE car would be. Without engine noise, even the base model sound system gives me clearer sounding music than in a ICE vehicle.

The most BEV manufactures insist on making "different" looking cars. They are more stylish, look better and new. The driver of electric cars is definitely "different" from others. While this is very subjective, but the "electric" technology is often associated with the care for planet and to the lot of other positive values. Consequently, a BEV owner is considered as a green oriented, young spirit, altruist person. All these positive values increase the self-esteem of BEV owners resulting in a willingness to possess BEV.

2 Specific NVH Issues for BEV

2.1 E-motor Noise

Permanent Magnet Synchronous Motor PMSM is used for most of electrified vehicle powertrains. PMSM rotor is equipped with permanent magnets and it follow rotating magnetic field. For the simplest "bi-polar" motor (one pair of poles) the rotation frequency is proportional to the stator current frequency. PMS motors with 4, 5 or 6 pole pairs equip the most electric powertrains. Higher number of poles implies proportional speed reduction for the same stator current frequency. The motor torque is proportional to the phase angle between the magnetic field and rotor – load angle. The sinusoidal current of varying frequency powering PMS motors is created using adequate electronic devices and PWM technique (Pulse Width Modulation). Pulse-width modulation (PWM), as it applies to motor control, is a way of delivering energy through a succession of pulses rather than a continuously varying (analog) signal. By increasing or decreasing pulse width, the controller regulates energy flow to the motor shaft.

Permanent magnet synchronous motor PMSM is used for most of electrified vehicle powertrains. The principal vibration excitation of PMS motors correspond to complex pattern of forces due to the magnetic flux and resulting Maxwell pressure. The forces change shape and intensity as function of time, while still being periodic. Both radial and tangential forces are to be accounted for, since both generate radial stator displacements. The Maxwell forces can be then developed into double Fourier series in time and space/angle domains and analyzed as the fundamental excitation components (Fig. 3).

Fig. 3. Principal sources of e-whining are: e-motor, reduction gears and power electronics.

Electrical drives are typically quieter than their ICE counterparts, but with very different NVH characteristics. While the ICE has a broadband noise spectrum, the electrical machine causes single tones in the audible spectrum, which unfortunately often corresponds to the frequency band in which the human ear is most sensitive. Therefore, despite their lower overall noise level, EVs may be sensed more unpleasant compared with ICE powered vehicles. Reducing the audible noise of the EVs asks for the analysis of the acoustic behavior including the electromagnetic force excitation, structural dynamic behavior and sound transmission and radiation, from the component (electrical machine) to the subsystem (electrical drive) and system level (powertrain).

Principal sources of whining are: electric motor, speed reduction gears and power electronics. Design parameters important for NVH are: number of poles, phases and slots, rotor and stator design, type of winding and skewing, stiffness and resonant frequencies, transmission error and axe rigidity, gear tooth surface and machining, frequency of PWM – Pulse Width Modulation and current control. All these parameters influence: the efficiency, the torque and the power characteristics of e-driveline. Consequently, a trade-off between NVH and other performances is generally needed.

2.2 Road Induced Noise and Vibration in BEV

Noise and vibration excited by the road are more noticeable in BEV because there is no masking effect of the corresponding IC engine excitation.

The tire/rolling noise of BEV is louder, since they are heavier due to the high battery mass. Such noise is not a emotional vector when driving but is pursued as a lack of acoustic insulation (quality). The tires impacts are at origin of the noise which is then can be attenuate (or amplified) by wheels, axles and overall vehicle.

The low-frequency vibrations are also produced by contact between the tires and the road. If correctly designed, the car body suspension filter this excitation. The optimization of powertrain suspension and stiffness of the BIW can improve the road induced vibrations (Fig. 4).

In order to isolate ICE idle excitation the modal frequencies of powertrain on elastic mounts have to be lower than the principal excitation frequency. For example, the principal engine order of 4-cylinder engine is 2 i.e twice of crankshaft rotational frequency. Moreover, the high road excitation at very low frequencies and engine order

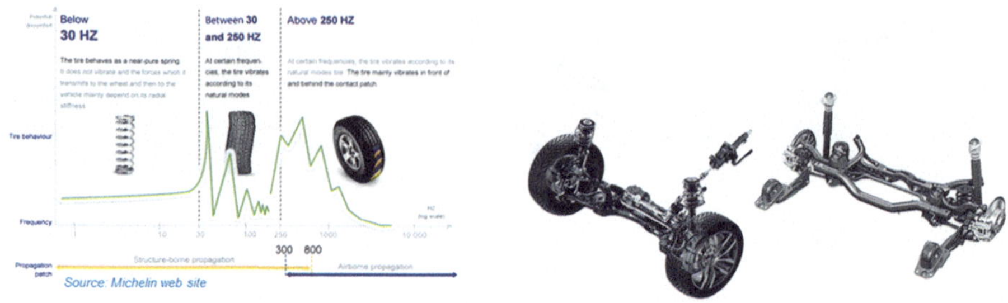

Fig. 4. Vehicle suspension behavior due to the road excitation for different frequency ranges.

0.5 characteristic for 2 time piston engines should be avoided, too. Consequently, all the resonances of powertrain on elastic mounts should be inside these two frequency limits.

When e-powertrain is considered, there is no idle speed excitation by principal engine order. Therefore, the modal frequencies of e-powertrain on elastic mounts can be higher in order to avoid the road excitation. Still, the mounts have to attenuate the structure borne part of whining excitation. For this purpose, a specific double filtering via a sub-frame may be used. There is no the specific, well suited, architecture of suspension to be used for e-powertrains. The design results from a trade-off between: the maximizing the mount spacing in order to diminish the saturation of mounts due to quasi-static load and the minimizing the part length in order to avoid resonances which may amplify the whining excitation (Fig. 5).

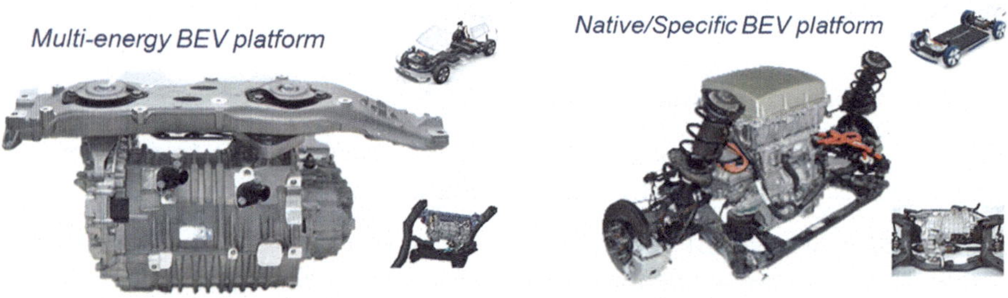

Fig. 5. Principal sources of e-whining are: e-motor, reduction gears and power electronics.

E-powertrain suspension system is platform dependent. The "multi-energy" platform, developed from an existing ICE one, have to allows integration of both IC and e-powertrains. Generally, a modified "pendulum" system is chosen for such platform, while for a "native" electric vehicle a dedicated mounting system can be designed.

In both cases, electric motor can be elastically mounted on an adequate sub-frame. The sub-frame is then fixed to the vehicle by the supplementary bushings, ensuring excellent isolation of whining structure borne noise (HF vibration).

The low rolling noise, underlines the quietness of BEV driving. When rolling, the road surface and tire tread pattern excite wheel over a wide frequency range. The wheel transmits part of this excitation to the vehicle and causes rolling noise inside of the vehicle. Depending on the design, the wheel can whether: isolate and attenuate the vibrations or amplify them if the wheel resonances are excited. The high rim rigidity is very important for a low road induced rolling noise. Specific design of alloy rims having high resonances is observed on BEV. Moreover, the road excitation can be amplified by car body deformations due to the low frequency modes of car body structure (torsion and flexion).

2.3 Battery Pack and BIW Rigidity

The capacity of Body-in-White (BIW) to attenuate road-induced noise and vibration is generally evaluated using two simple criteria: static torsional stiffness and frequency of first torsional mode. Same type of engineering criterion can be established for flexural deformation.

The structural rigidity of multi-energy platform highly depends on battery pack stiffness and mass. As a general rule, battery pack increases the static stiffness due to the additional rigidity fixed to the structure but diminishes the principal torsional mode frequency due to the added mass. Nevertheless, if well designed, the multi-energy platform preserves an equal performance attenuation of road excitation in both versions ICEV and BEV (Fig. 6).

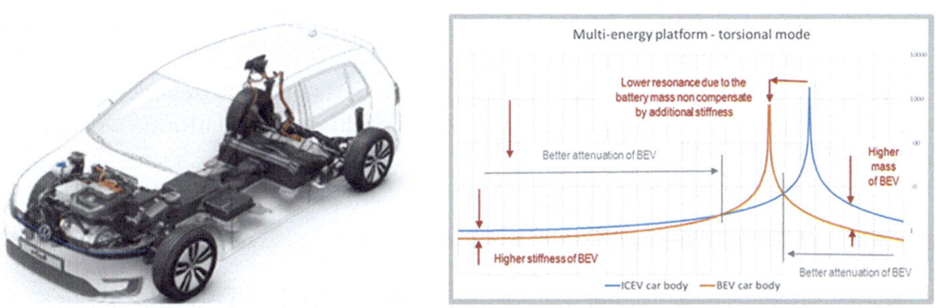

Fig. 6. Battery pack fixed to the BIW increases the stiffness and decreases the resonant frequencies (torsion and flexion) due to the added mass.

Native BEV platforms maximize battery packaging and autonomy. As a rule, native platforms, compared to the multi-energy ones, have higher static torsional rigidity due to the battery pack stiffness and lower torsional frequency due to battery mass. An appropriate design of both, platform and battery-pack, can improve the attenuation of structure borne noise and vibration induced by road up to 4–5 dB in the frequency range of 50–500 Hz.

2.4 Acoustic Transparency and Insulation

Both, external traffic noise and driving generated noise are to be accounted for when designing NVH treatment of BEV. The lack of IC engine masking, increase the importance of insulation from the external noise. The noise of surrounding traffic is very penalizing for the customer acoustic comfort and overall impression of quality and market positioning of vehicle. Therefore, the vehicle have to be soundproof up to certain sound pressure level, which can be achieved by an appropriate (high) thickness of windshield and car glass, by sealing of car doors (no acoustic leakage) and by applying absorption material in the driver's compartment (Fig. 7).

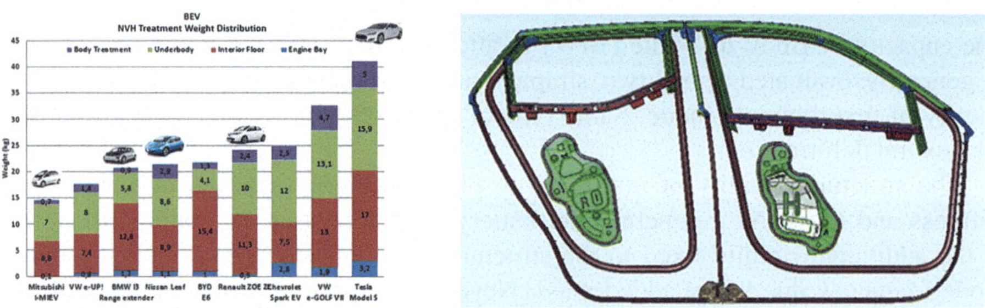

Fig. 7. A performant insulation is important for BEV. Both external noise and nose generated by the vehicle have to be taken into account when design the acoustic treatment. The sealing in order to diminish acoustic leakage and corresponding transparency due to external noise.

During the benchmarking it is observed that the mass of NVH treatments in ICEV and BEV is approximatively the same, at least on multi-energy platforms where comparison is easy to do.

2.5 Noise of Auxiliaries and Other Devices

In the ICEV the IC engine drives accessories. In the BEV the different devices as compressors, pumps and fans are to be electrically driven. The noise of such devices is particular important, since exacerbated, without IC engine masking. Such a new type of accessory generates new type of NVH problems (whining, electric whistling etc) (Fig. 8).

The noise generated by the electrified accessory in not the only acoustic issue. Disturbing dysfunctional BSR noises (buzz, squeak and rattle) can result in a very low car positioning with respect to the perceived quality. Consequently, the NVH specification on system & device noise have to be strengthened, in order to take into account relative quietness of BEVs.

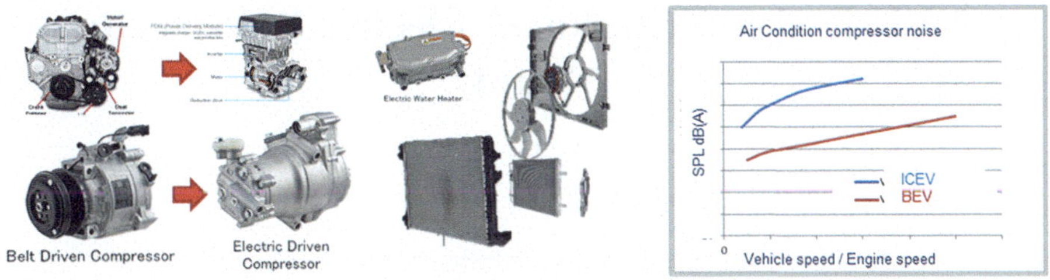

Fig. 8. BEV accessory is electrified and potentially represents high whining risk. Moreover, the lack of ICE engine masking results in more sever specifications for most of driving conditions.

3 State of the Art and Concluding Remarks

Battery Electric Vehicles (BEVs), compared to classic internal combustion engine (ICE) vehicles, are fairly simple and easy to operate. The simplest powertrain architecture consists in a high voltage battery, an electric motor with power electronics controller and a single speed gearbox.

With zero tailpipe emissions, BEVs also offer a clean alternative to vehicles with ICEs by helping to reduce exposure to air pollution resulting from fuel combustion and limiting external noise. This is especially relevant in urban areas and along major transportation axes. Electric cars eliminate engine noise, and the relatively high-pitched noise electric motors do emit does not propagate as far. However, the faster vehicles are traveling, the more road and tire noise dominate, and the more wind passage noise contributes; those two are particularly important for internal acoustic comfort. Even for low speed urban driving conditions, BEV can have high tire noise contribution due to heavy battery and higher tire pressure.

Since without IC engine, the electric powertrain offers favorable conditions for a low operating external noise, still a careful NVH integration is necessary to achieve a NVH behavior compatible with an idea of quiet electric driving important for BEV customer.

Reference

1. Gavric, L.: Electric vehicles and principal NVH issues: Proceedings of Automotive NVH Comfort – Le Mans, France 17–18 October 2018; SIA Conference (2018).

NVH Design Criterion of Electric Powertrain Mounts

Qiang Kang and Perry Gu(✉)

Geely Automobile Research Institute, Ningbo, China
gupengyun@geely.com

Abstract. Compared with the mechanical and combustion noise of low-frequency firing orders of an internal combustion engine, the electric drive assembly noise of electric vehicles is mainly the high-frequency whining noise generated by electromagnetic forces and transmission gear meshing. Although the radiated sound power is far less than that of an internal combustion engine, the high frequency noise of the electric motor and the transmission is much annoying. Noise below 2000–2500 Hz generated by an electric drive assembly in a vehicle passenger compartment is mainly contributed by structural transfer paths. Due to the high torque of the electric drive motor of electric vehicles, it is a challenge to achieve better mount isolation performance at the same time when the anti-torque performance of the electric drive assembly mounting system is realized. In this paper, by comparing the vibration characteristics of electric vehicles and traditional ICE vehicles, comparing different mount layout schemes, and analyzing the main factors affecting the vibration isolation rate of the electric drive assembly mounting system, the general design principles of the electric drive assembly mounting system are summarized to achieve balanced attribute performance.

1 Introduction

With the development of energy saving, environmental protection and intelligentization of vehicles, the automobile industry is moving toward electrification, automated driving and shared mobility. NVH of automobiles is also facing new challenges brought by electrification. Compared with the low-frequency mechanical and combustion noise of an internal combustion engine, the electric drive assembly noise is mainly the high-frequency whining noise generated by the electromagnetic force of the motor and gear meshing of the reducer [1]. Although the radiated sound power of the high frequency noise is much lower than that of the internal combustion engine noise, the high frequency noise of the motor and the reducer is subjectively quite annoying. In order to meet the user's expectation of quietness in vehicle interiors, it is necessary to completely prevent the whining noise of the electric drive assembly from transmitting to the vehicle interior. On one hand, the high-frequency whining noise of the motor and reducer heard by the passengers in the interior is transmitted from the motor cabin to the interior by air-propagation paths; On the other hand, the whining noise is transmitted by structural propagation paths such as mounts and vehicle body structures. Studies have shown that the noise below 2000–2500 Hz frequency

W. Siebenpfeiffer (Hrsg.): *Automotive Acoustics Conference 2019*, Proceedings, S. 10–27, 2020. https://doi.org/10.1007/978-3-658-27669-0_2

of the electric drive assembly in the interior is mainly contributed by the structural propagation path [2]. Therefore, the vibration isolation design of an electric drive mounting system is very important for controlling the whining noise in a vehicle. The mounting system needs to be able to resist the torsion of the assembly on one hand and isolate the vibration of the assembly transmitted to the body structure on the other hand. Due to the large torque of the electric vehicle's drive motor, it is a challenge to improve the vibration isolation performance of the electric drive assembly mounting system as much as possible while achieving the torsion resistance.

Xu zhongming et al. optimized the mounts of electric vehicles [3], but still used the decoupling rate commonly used in IC engine mounts as a target, which was obviously inappropriate. This is because the excitation frequency of an electric drive motor is completely different from that of the engine. The motor excitation frequency is much higher than the rigid body modal frequency of the electric drive assembly, and the excitation force is very small, so the electric drive assembly will not resonate in general. Li yaoyao et al. took the minimum dynamic reaction force at the mounting support as the optimization objective [4], but they only analyzed and tested up to 100 Hz, which was obviously far lower than the high-frequency excitation and noise frequency range of the electric drive assembly.

This paper systematically analyzes vibration characteristics of different mount layout schemes, and main factors affecting the vibration isolation rate of the electric drive assembly. The general principles of the design of the electric drive assembly mounting system based on the two functional requirements of torsion resistance and vibration isolation are summarized.

2 Comparison of Vibrational Characteristics of Powertrain in Electric and ICE Vehicles

Table 1 lists powertrain mass and mount stiffness of an ICE vehicle, and lists rigid body modes of the powertrain of two ICE vehicles. Table 2 lists powertrain mass and mount stiffness of a pure electric vehicle, and lists powertrain rigid body modes of five electric vehicles.

Table 1. Powertrain data of traditional ICE cars

Powertrain mass（1.0TD-6DCT）： 188Kg			
Mount stiffness(N/mm)			
	Ku	Kv	Kw
Engine mount	-	-	129
Transmission mount	-	-	121
Rear mount	143	-	-
Powertrain rigid mode frequency（Hz）			
	Geely	Ford	
Lateral	4.6	8.6	
Bounce	7.1	9.9	
For-aft	7.3	7.9	
Yaw	12.9	13.8	
Roll	15.1	19.3	
Pitch	9.9	9.1	

Table 2. Powertrain data of electric cars

Powertrain mass（motor + reducer）：92.5kg					
Mount stiffness(N/mm)					
	Ku		Kv		Kw
Engine mount	194		201		268
Transmission mount	194		201		268
Rear mount	233		10		10
Powertrain rigid mode frequency（Hz）					
	Geely	BOLT	LEAF	I3	Model X
Lateral	4.9	12.8	4.4	9.4	21.3
Bounce	16.2	13.8	6.0	12.2	9.6
For-aft	20.9	14.7	20.9	13.2	11.7
Yaw	16.3	22.6	12.8	26.9	47.7
Roll	28.4	24.2	20.9	37.2	39.4
Pitch	35.0	26.7	29.3	23.2	32.0

By comparing Tables 1 and 2, it can be seen that the weight of the electric drive assembly is about half that of the traditional ICE powertrain, and the mount stiffness is about twice that of the traditional ICE powertrain. Therefore, the rigid body modal frequency of the traditional vehicle powertrain is generally designed between 5–20 Hz, while the rigid body modal frequency of the electric drive assembly is 25–50 Hz, which is about 4 times of that of the traditional vehicle.

Compared with the traditional powertrain, the torque of the electric drive assembly is relatively large, but the fluctuation is very small. Furthermore, the excitation of the electric drive assembly mainly comes from the high frequency electromagnetic force and gear meshing force, which is far higher than the electric drive assembly rigid body modes. Therefore, although the powertrain rigid body mode of the electric vehicle is about 4 times that of the traditional vehicle, it is still far less than the excitation frequency.

Figure 1 shows the vibration colormap of the active side of the rear mount of the electric drive assembly under the operation of full acceleration of an electric vehicle. As it can be seen, the frequency of its main excitation is between 400 and 8000 Hz.

In addition to the vibration isolation performance of the mounting system, the torsion resistance performance of the mounting system should also be considered. Especially for electric vehicles, the motor torque at low rpms is much large than the torque of typical ICEs, the response is fast, and there is a strong energy recovery torque when the vehicle is coasting. Therefore, the transient impact on the vehicle is greater, and it is easy to cause the powertrain shake in tip in/out conditions. Table 3 shows subjective evaluation scores of noise and shake in an electric vehicle with different mount stiffness schemes. Among these three schemes, scheme A is to increase the left and right mount stiffness from 260 N/mm to 360 N/mm, and scheme B is to increase the rear mount stiffness from 230 N/mm to 290 N/mm in addition to scheme A. It can be seen that there is a different performance trend between reducing the noise and reducing the shake in the car.

The subjective scores suggested that in order to reduce the shake, the mount stiffness should be increased. However, the increase of mount stiffness leads to the decrease of mount isolation rate, which results in the increase of the interior noise.

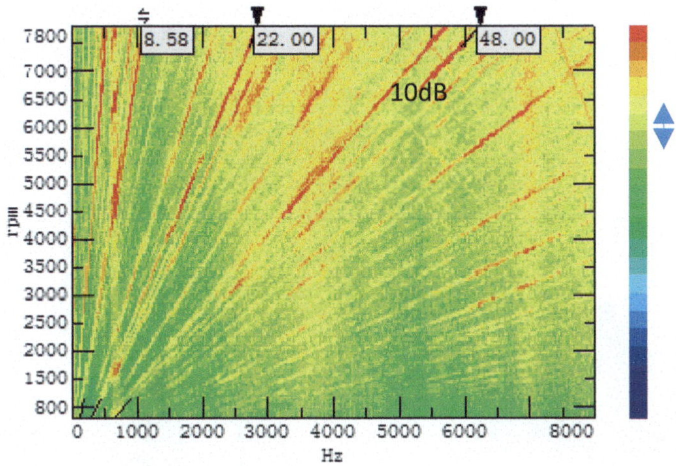

Fig. 1. Color map of the active side vibration of an electric vehicle powertrain mount (WOT)

Table 3. Subjective evaluation of noise and shake in an electric vehicle with different mount stiffness schemes

Scheme	Subjective grading	
	Interior whining	Tip in/out shake
Original scheme	6.75	6
Scheme A	6.5	6.5
Scheme B	6	7

Therefore, based on above analysis, the mounting system of electric vehicles should first meet the requirement of no shake (that means minimizing the attitude change of the powertrain). On this basis, the mount stiffness should be reduced as far as possible to improve the vibration isolation rate of the mount.

3 Comparative Analysis of Mounting Layout

For a three-point mounting layout, currently there are generally five forms of powertrain mounting layout for electric vehicles, as shown in Fig. 2.

The above five layout forms are simplified to theoretical models, as shown in Fig. 3. Same inertia and torque of the powertrain and same mount stiffness curves are used in these five models.

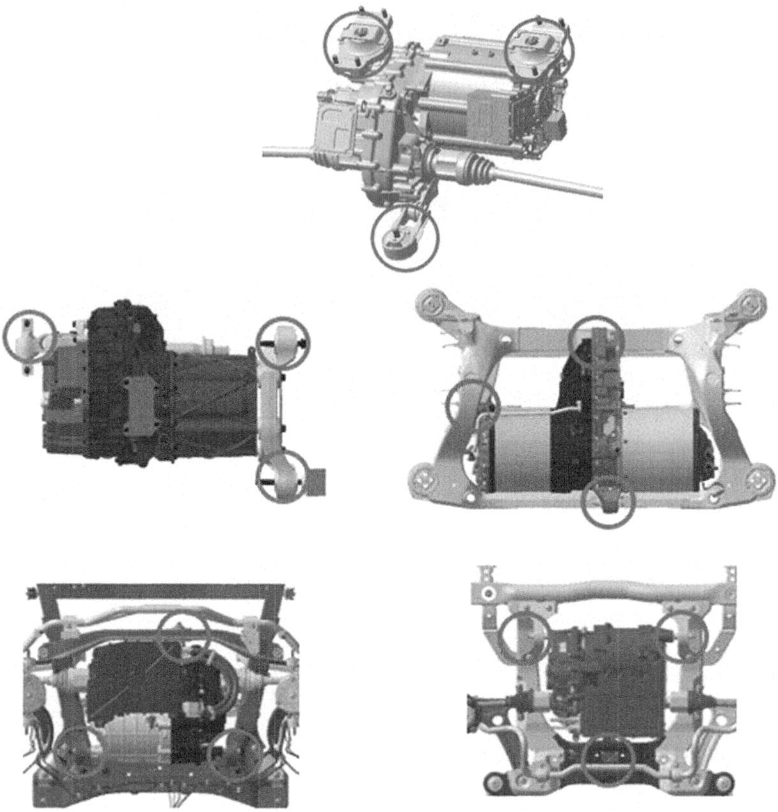

Fig. 2. Mounting layouts of powertrain of electric vehicles

The attitude changes (mainly considering the rotation angle around the Y axis) of the powertrain and the mout force with various layout schemes under different conditions of 100% torque, 50% torque and 30% torque are analyzed, respectively. The results are shown in Figs. 4 and 5. It can be seen from the analysis results that:

- The attitude change and mount force are the largest in TRA arrangement. Its rotation angle and force are about 2 times of other layout. This will require the mount stiffness to increase and result in decrease of the mount isolation rate. Therefore, the TRA axis arrangement method commonly used in tranditional vehicles cannot be used for the mounting layout of electric vehicles.
- There is little difference of the powertrain attitude change and the total force of mounts under the three layouts of front 2+ rear 1, left 2+ right 1 and left 1+ right 2. But the mount force is uneven among the three mounts under the layouts of left 2+ right 1 and left 1+ right 2. It is seen that one of the mount force is very small under these two layouts.

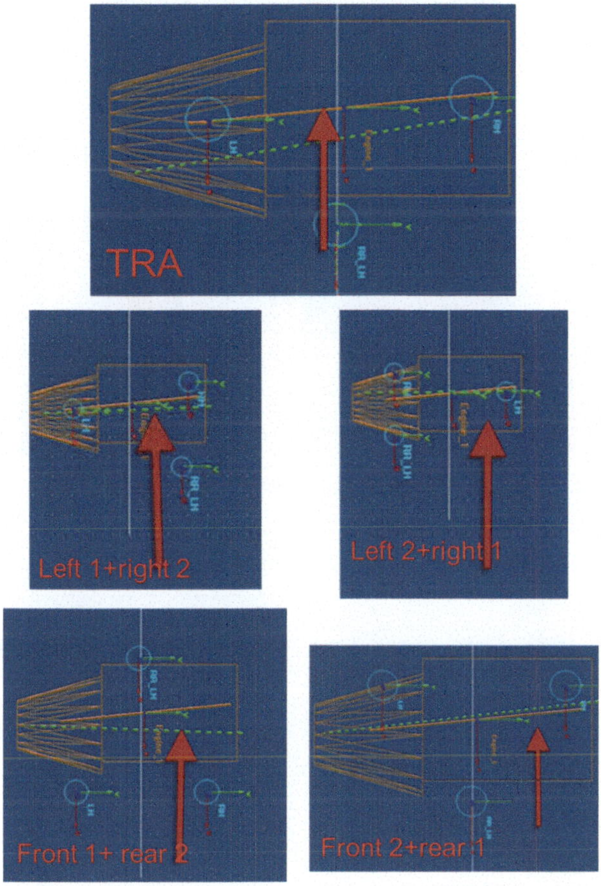

Fig. 3. Analysis model for powertrain attitude change and mount force

- The optimal layout is front 1+ rear 2. In this layout, the attitude change and total force are the minimum, because the mounting points are at the maximum distance from the TRA axis.

It is summarized that the mount layout of electric vehicle cannot follow the TRA axis arrangement method of traditional ICE vehicles, but should keep the elastic center of each mount as far as possible away from the TRA axis, so as to minimize the mount force and the attitude change of powertrain. Meanwhile, because the mount force is reduced, mounts with smaller stiffness can be used in order to improve the isolation rate, while keeping the attitude change unaffected.

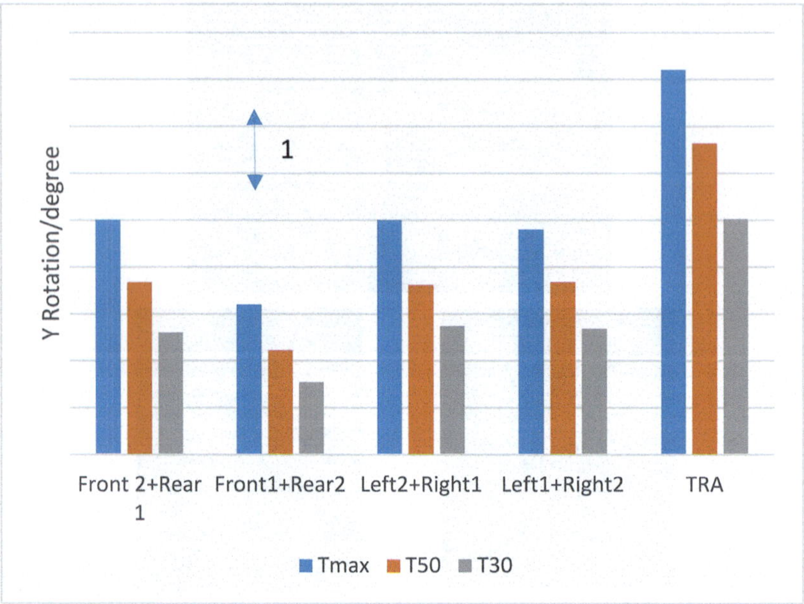

Fig. 4. Comparison of attitude changes of electric drive assembly under different mounting layouts

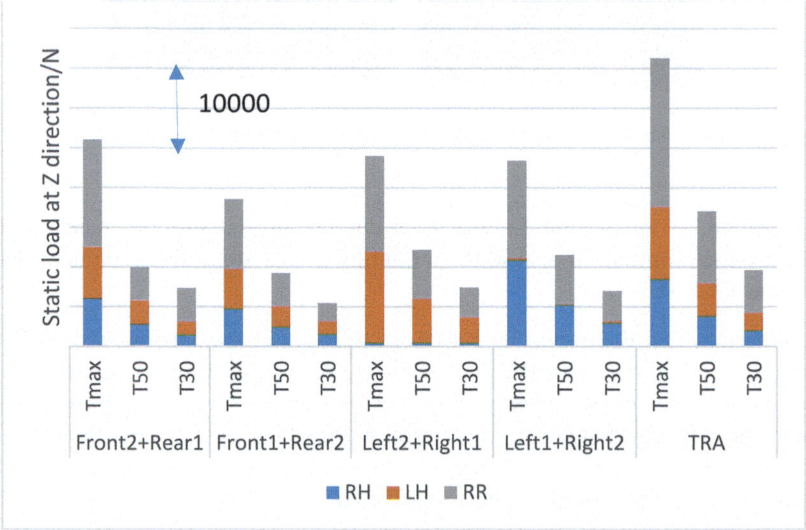

Fig. 5. Comparison of mounting forces of electric drive assembly under different mounting layouts

4 Vibration Isolation Analysis of Mounting System

4.1 Mounting Isolation Test

Microphones are placed in the vehicle's motor cabin and at the position of the driver's ears in the vehicle interior. Three-direction acceleration sensors are arranged on the mounts' active side and passive side bracket. The vehicle was placed on a chassis dynamometer in a semi-anechoic room. The noise and vibration is measured under the conditions of full load acceleration and coasting energy recuperation. The vibration isolation test results in three directions of each mount are shown in Fig. 6. Order 8.58 and 22 are the main order of output gear noise and input gear noise of the reducer respectively, and order 48 is the main order of motor noise.

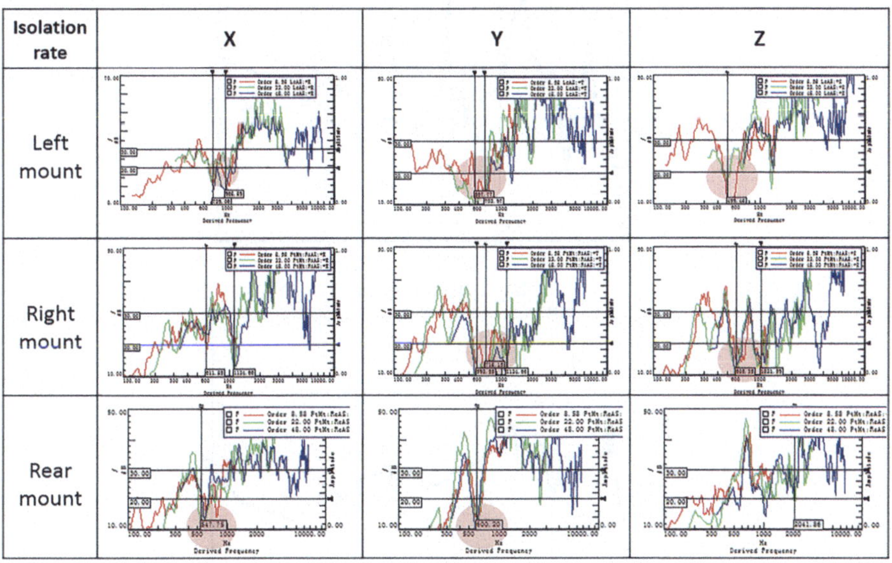

Fig. 6. Mounting isolation rate (red: 8.58 order; green: 22 order; blue: 48 order)

4.2 The Effect of Mount Bracket Mode on Vibration Isolation Rate

As it is seen from Fig. 6, the mount isolation rate decreased significantly in the range of 500–1000 Hz. This is because the mount bracket has resonant modes within the range of 500–1000 Hz, as shown in Fig. 7 (the valley frequency of the dynamic stiffness curve of the bracket in the figure is the modal frequency of the bracket). The valley frequencies of vibration isolation rate and the resonant frequencies of the bracket are listed in Table 4. It can be seen that the frequencies are almost identical, indicating that the low vibration isolation rate is mainly due to the resonant response of the bracket.

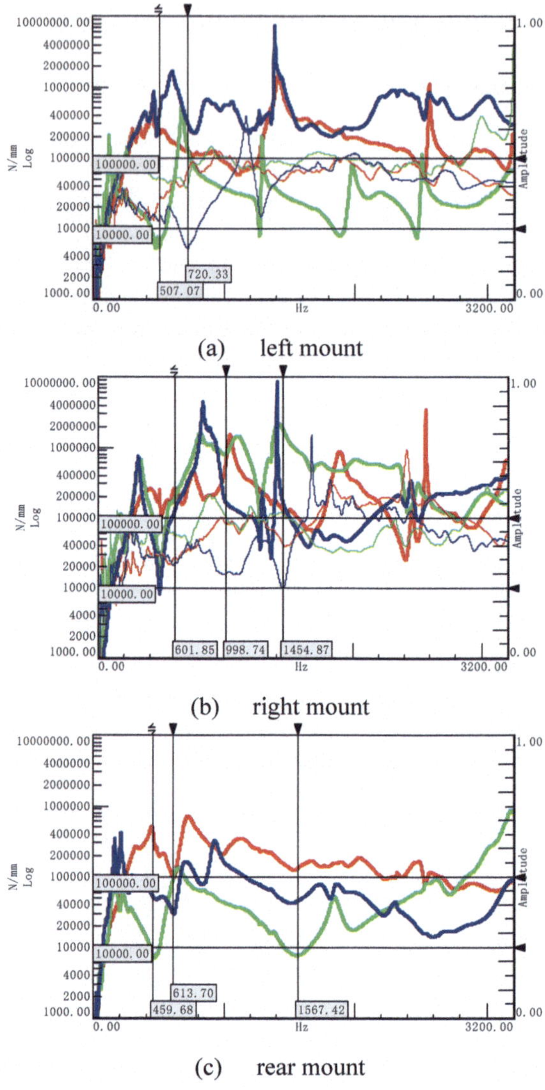

Fig. 7. Dynamic stiffness curve of mount bracket (red: x; Green: y; Blue: z; Thick line: active side; Thin line: passive side)

Table 4. Valley frequencies of vibration isolation rate and bracket resonance frequencies

	Valley frequency of isolation rate (Hz)	Bracket resonance frequency (Hz)
Left	500	507
	729	720
Right	606	601
	1021	998
Rear	600	613

Figure 8 shows the order curve of vehicle interior noise obtained from the test. It can be seen that below 1500 Hz, the peak frequency of the reducer gear and motor order noise in the car is consistent with the resonance frequency of the bracket. It is concluded that the resonance characteristics of the mount bracket are important influencing factors for the noise of the electric drive assembly transmitting into the cab at low speed (below about 40 km/h).

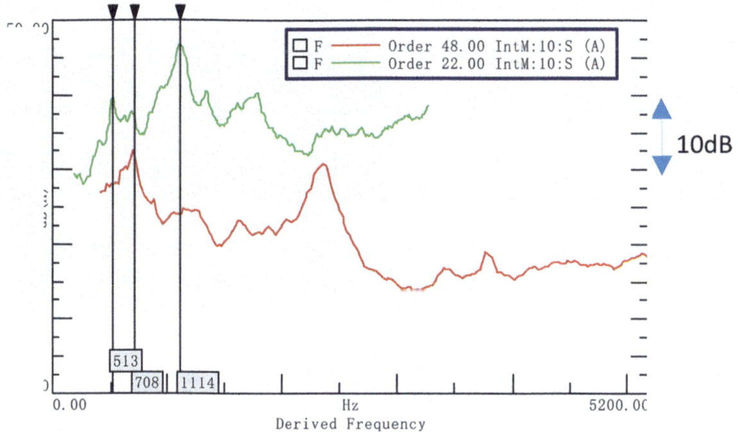

Fig. 8. Interior noise order curve

4.3 Effect of Mount Rubber Stiffness on Vibration Isolation Rate

For the mounting system, its vibration isolation rate can be theoretically calculated by:

$$\mathrm{IR} = 20\lg\frac{1/\mathrm{K}_p + 1/\mathrm{K}_a + 1/\mathrm{K}_i}{1/\mathrm{K}_p + 1/\mathrm{K}_a} \tag{1}$$

Where, IR is the isolation rate, the unit is dB, Kp is the stiffness of the passive side, Ka is the stiffness of the active side, and Ki is the stiffness of the isolation element.

According to formula (1), isolation rates can be listed as K_a/K_i and K_p/K_i at different values, as shown in Table 5.

As shown in Table 5, in order to satisfy the vibration isolation rate at least greater than 20 dB (the part with green color in the table), the stiffness ratio between the active and passive sides and the vibration isolation element needs to be more than 20.

Table 5. Vibration isolation rate list

K_p/K_i \ K_a/K_i	1	5	10	20	30	40	50	100
1	3.52	5.26	5.62	5.81	5.88	5.91	5.94	5.98
5	5.26	10.88	12.74	13.98	14.46	14.72	14.88	15.21
10	5.62	12.74	15.56	17.69	18.59	19.08	19.40	20.08
20	5.81	13.98	17.69	20.83	22.28	23.13	23.69	24.94
30	5.88	14.46	18.59	22.28	24.08	25.17	25.91	27.63
40	5.91	14.72	19.08	23.13	25.17	26.44	27.32	29.42
50	5.94	14.88	19.40	23.69	25.91	27.32	28.30	30.71
100	5.98	15.21	20.08	24.94	27.63	29.42	30.71	34.15

The dynamic stiffness of the mount and the bracket active and passive side obtained by the test is shown in Fig. 9. Among them, the mount stiffness can only be measured up to 1000 Hz, since it is restricted by the capability of the test bench. The preload on the mount during the test is the gravity of the powertrain that is supported by the mount.

As shown in Fig. 9, the dynamic stiffness of left mount increases sharply at high frequency, with a dynamic to static ratio of 1.4 at 100 Hz and 17.5 at 1000 Hz. The dynamic stiffness of the rear mount changes little at high frequency, with a dynamic to static ratio of 2.5 both at 100 and 1000 Hz. Due to the increase of the left mount stiffness, the ratio between the dynamic stiffness of the bracket passive side and that of the left mount at 1000 Hz is only 5, which will greatly reduce the vibration isolation rate of the mount.

In addition, it can be seen from the dynamic stiffness curve of the mount rubber that there is a resonance peak, which is around 600 Hz for left mount and around 400 Hz for rear mount. This is because at high frequency (above 100 Hz), the rubber vibration isolation element does not conform to the massless assumption, but has the characteristic of distributed mass, so standing wave effect should be considered [5]. The participation effect of the dynamic mass of the mount rubber cannot be ignored for the high frequency excitation (400–8000 Hz) generated by the electric drive assembly.

Therefore, in order to improve the vibration isolation rate of the system, it is necessary to avoid coupling of the mode frequency of mount bracket with the internal resonance frequency of mount rubber, and mount rubber with different parameters should be adopted to improve the vibration isolation rate of the system [5].

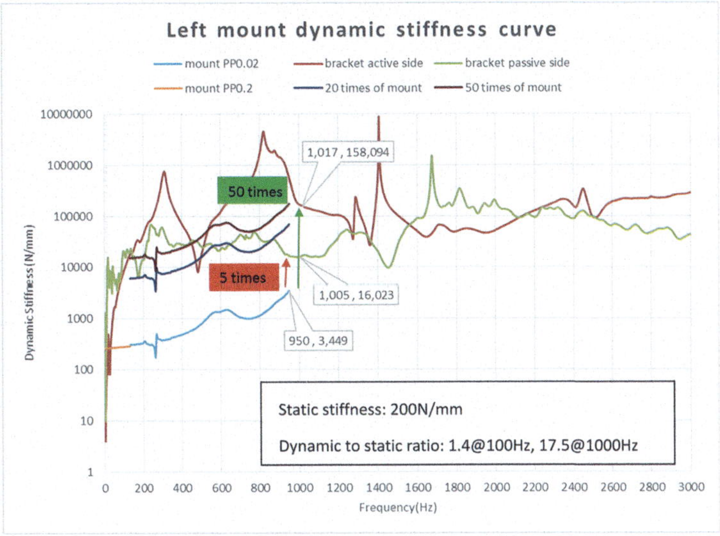

(a) left mount

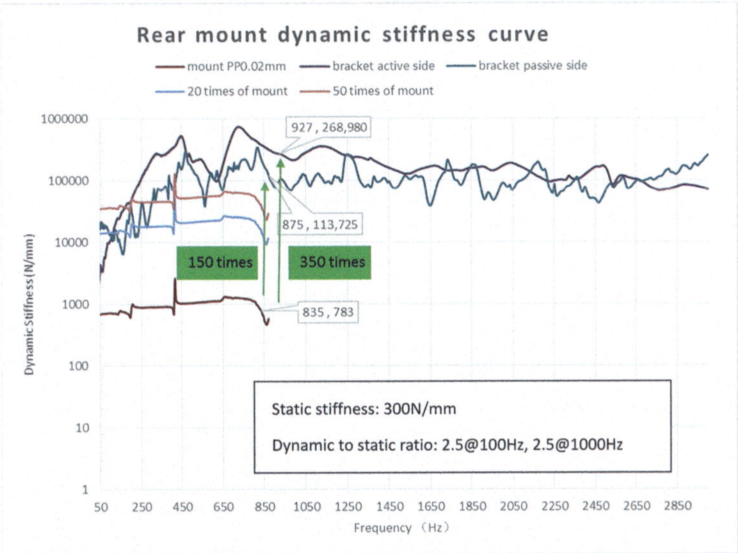

(b) rear mount

Fig. 9. Dynamic stiffness curves of mount and bracket

The mount stiffness in Fig. 9 is the test result under the condition that the preloaded is the gravity of the powertrain. This test condition can be equivalent to the operation condition that the powertrain is under a small load (small pedal acceleration, constant speed cruising, weak energy recuperation). For the powertrain noise of electric vehicles, the other key conditions to be considered are medium and heavy load conditions (half pedal acceleration, full pedal acceleration, and strong energy recuperation). Being different from traditional ICE vehicles, because the motor torque of electric vehicle is generally very large, full pedal acceleration condition is basically the worst working condition in the mount 28 working conditions. At this condition, the mount is in a fully compressed state. As shown in Fig. 10, when fully compressed, the static stiffness increases to a maximum of 18 times as compared with that under small load, and the dynamic stiffness will be also significantly higher than that measured in Fig. 9. Therefore, to

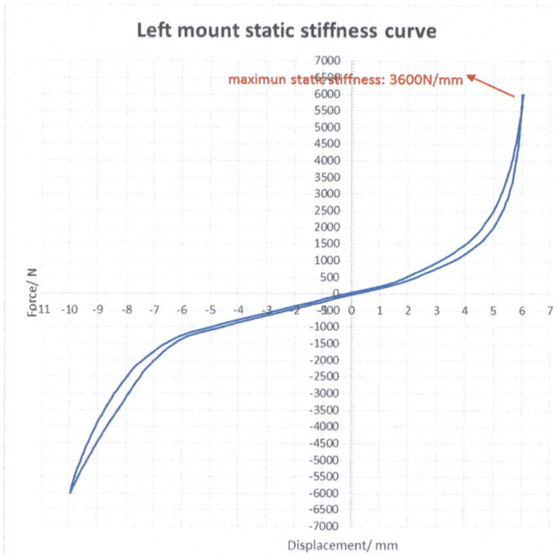

(a) Left mount

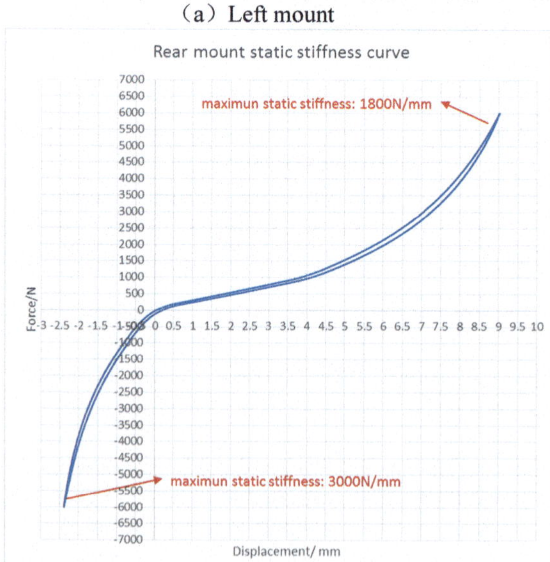

(b) Rear mount

Fig. 10. Mount force-displacement curve

increase the mount vibration isolation rate at high frequency, it needs to minimize the mount static stiffness as much as possible, as well to lower the dynamic to static ratio at high frequency under interested key operating conditions (mount preload is powertrain gravity, mount preload is load on the mount under condition of strong energy recuperation, mount preload is load on the mount under condition of half pedal acceleration, mount preload is load on the mount under condition of full pedal acceleration).

4.4 Influence of Soft Connection of Sub-frame on Vibration Isolation Rate

The mode of mount bracket reduces the vibration isolation rate of mount, and the excessive high dynamic stiffness of mount is not conducive to the improvement of vibration isolation rate. In order to improve the vibration isolation rate, another feasible method in engineering is to use double layer vibration isolation, such as soft connecting a sub-frame to a car body through bushings. The transmission curve of double layer vibration isolation is shown in Fig. 11. With two-layer isolation, the decline slope of the transfer curve after the resonant region increases from 12 dB/oct in the single-layer system to 24 dB/oct, which is conducive to the isolation of high-frequency vibration.

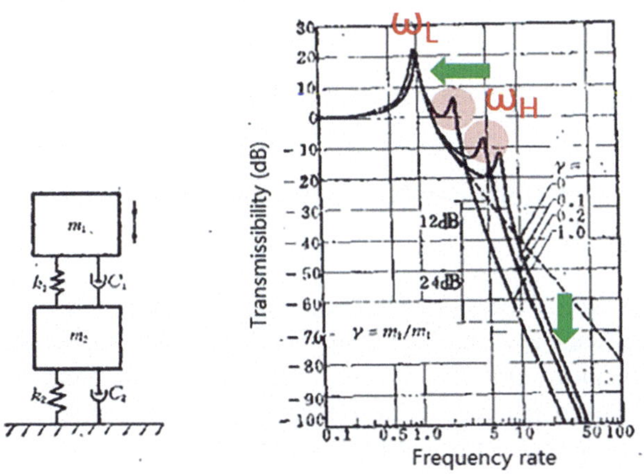

Fig. 11. Double isolation system and its transmissibility curve

Figure 12 shows the comparison between the vibration isolation rate curve after adopting a sub-frame with soft connection and the vibration isolation rate curve of the single-layer mounting. As shown in the figure, due to the influence of the mode of the mounting bracket, the vibration isolation rate of the mounting below 200–600 Hz is very low. After using the sub-frame with soft bushings, it can compensate for the insufficient mount vibration isolation rate. Generally, the vibration isolation rate of the active end of the mount to the body can be improved by more than 15 dB.

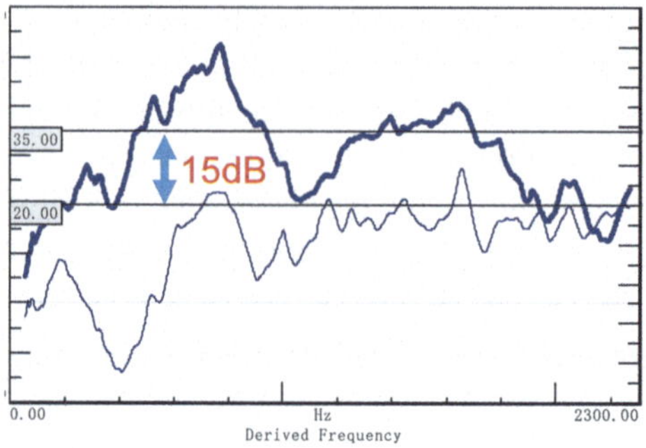

(a) Left mount

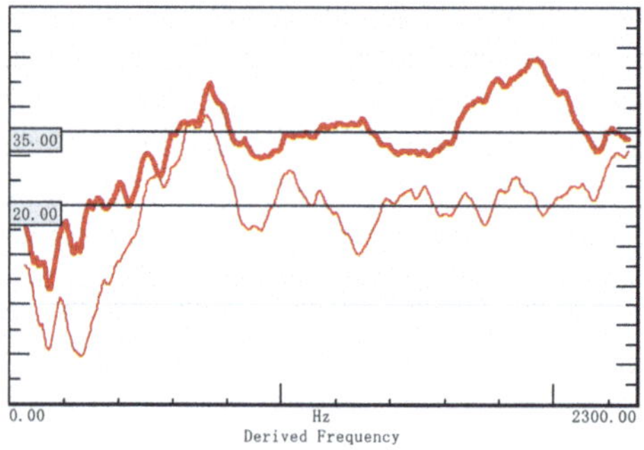

(b) Rear mount

Fig. 12. Vibration isolation rate curve (thick line: double isolation; thin line: single isolation)

5 Comparison of Reducer Noise in the Vehicle and on the Test Bench

In order to directly verify the influence of mounting bracket on the noise of a reducer, the electric drive assembly and mounting system were removed from the car and installed on a bench for testing, as shown in Fig. 13. After the first testing, the mounting system is removed and the reducer is rigidly connected to another test bench (without mount bracket), as shown in Fig. 14.

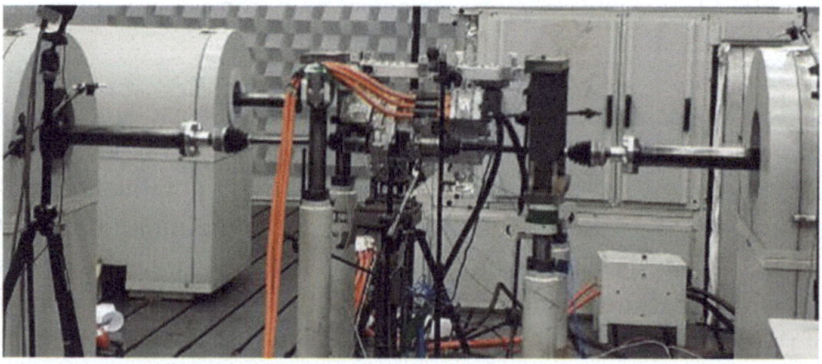

Fig. 13. NVH test of the electric drive assembly

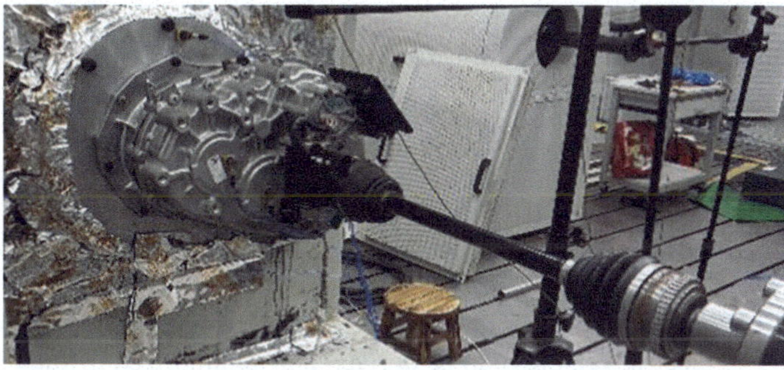

Fig. 14. NVH test of the reducer

A comparison is made among the 22th order reducer near-field noise tested in the car, on the powertrain test bench and on the reducer only test bench, as shown in Fig. 15. As shown in the figure, in the range of 500–1000 Hz, the 22th order noise of the reducer in the vehicle is significantly higher than the test result on the bench. The

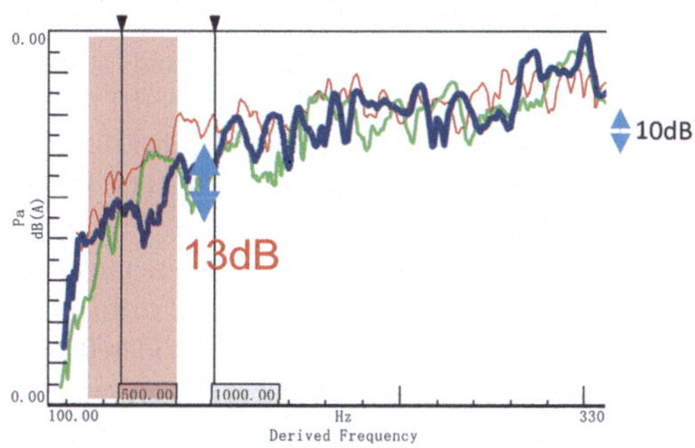

(a) WOT operating condition

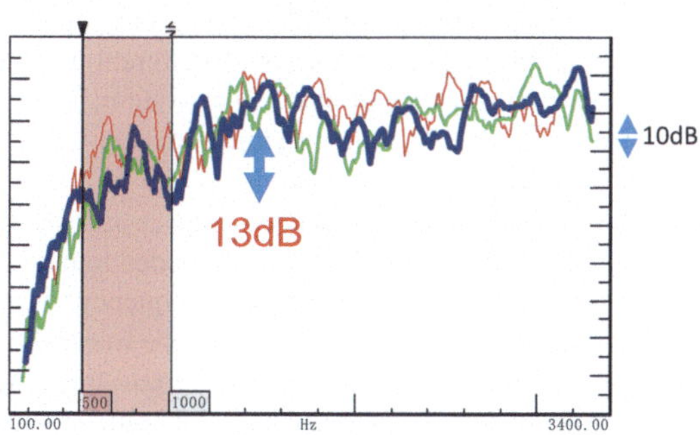

(b) Regenerative energy recuperation during coasting

Fig. 15. Comparison of reducer gear order noise (red: in the vehicle; Green: on the assembly bench; Blue: on the reducer bench)

noise in the vehicle is about 13 dB higher than that on the bench. Thus, it can be seen that the mount bracket mode has a great influence on the noise of the reducer transmitted to the vehicle, which needs to be improved.

6 General Design Principles for Mounting of Electric Drive Assembly

Based on above analysis, the general design principles for the mounting of electric drive assembly of electric vehicles can be derived and summarized in Table 6:

Table 6. Design principles for mounting of electric drive assembly

Design principles	
Mounting layout	• cannot use TRA layout; • the center of the mount should be as far away from the TRA axis as possible to achieve maximum torsion resistance; Meanwhile, the maximum static stiffness of the mounts is minimized • reasonably arrange the mount position to make the load of each mount even
Rigid body modes and decoupling rates	• rigid body modal frequency avoids suspension hop/tramp frequency and steering wheel modal frequency; • pitch decoupling rate >80%, other direction not required
Vibration isolation rate	• vibration isolation rate of 400–8000 Hz should be greater than 30 dB; • in order to meet the isolation rate under main operating driving conditions, the mount dynamic stiffness under the four operating conditions should be considered and tested up to at least 1000 Hz: mount preload is powertrain gravity, mount preload is load on the mount under condition of strong energy recuperation, mount preload is load on the mount under condition of half pedal acceleration, and mount preload is load on the mount under condition of full pedal acceleration; • adopting mount rubber with different parameters to improve the vibration isolation rate of the system; • adopting double layer vibration isolation (soft connection of subframe)
Mount bracket	• mount bracket should be compact in design and the higher the mode, the better. It is recommended to be greater than 1000 Hz • make the internal resonance frequency of the mount rubber as small as possible, and separate the internal resonance frequencies of different mounts with certain frequency intervals • avoid coupling the mount bracket mode with the internal resonance frequency of the mount rubber • if the mounting bracket mode does not meet the requirements, a damper shall be added to the bracket as required

7 Summary

By comparing the powertrain characteristics of electric vehicles and traditional ICE vehicles, it is concluded that the electric vehicle mounting system should first meet the requirement that the vehicle is without shake. Based on this pre-requisite condition, the mount stiffness should be reduced as much as possible to improve the vibration isolation ratio of the mounting system. By comparing the different mounting layouts of the electric drive assembly, it is concluded that the electric vehicle mounting layout cannot follow the TRA axis layout idea of the conventional ICE vehicle, but the mounting elastic center points should be kept as far as possible from the TRA axis, thereby minimize the force on the mounts and minimize the change in attitude of the powertrain.

Furthermore, by analyzing the mounting vibration isolation rate, it is concluded that the resonance characteristics of the mounting bracket are important factors for the electric drive assembly noise in the vehicle at low speed (about 40 km/h or less). The mounting bracket needs to be compact in design, and the higher the modal frequency, the better the mounting system isolates the vibration. It is recommended to be greater than 1000 Hz, and avoid the coupling of the mounting bracket modal frequency and the internal resonance frequency of the mounting rubber.

Finally, in order to improve the high-frequency vibration isolation rate of the mounting system, it needs to minimize the static stiffness of the mounting under the concerning critical conditions as much as possible, and to lower the dynamic to static ratio at high frequencies. It is also found that a soft-connected sub-frame can compensate for the insufficient mounting vibration isolation rate, and the vibration isolation rate of the mounting active side to the vehicle body is increased by more than 15 dB.

References

1. Ma Cong-gan, Zuo Shu-guang, et al. Order feature analysis of electromagnetic noise in a permanent magnet synchronous motor considering current harmonics [J]. Journal of shock and vibration, 2014(15):108–113.
2. David Lennström, Roger Johnsson, Arne Nykänen, etc. Determination of Radiated Sound Power from an Electric Rear Axle Drive In-Situ and its Contribution to Interior Noise [J]. SAE Technic Paper 2013-01-9120.
3. Xu Zhongming, Li Xiao, et al. Optimization of the powertrain mounting system of electric vehicle [J]. Automotive Engineering, 2012, 34(9):806–810.
4. Li Yaoyao, Wang Hui. Optimization of powertrain mount system of electric vehicle [J]. Journal of Chongqing jiaotong university (natural science), 2015, 34(1):135–139.
5. Zeng, X., Liette, J., Noll, S., and Singh, R., Analysis of Motor Vibration Isolation System with Focus on Mount Resonances for Application to Electric Vehicles [J]. SAE Int. J. Alt. Power. 4(2):2015.

Methods and Solutions for the Integration of the HV Battery into the NVH Development of a BEV

Howard Brentnall[1], Francesca Ronzio[2](✉), Federico Di Marco[2], and Loic Wang[3]

[1] Jaguar Land Rover, Banbury Road, Gaydon, Warwick CV35 0RR, UK
[2] Autoneum Management AG, Schlosstalstrasse 43, 8406 Winterthur, Switzerland
francesca.ronzio@autoneum.com
[3] Autoneum France SASU, Rue des Chevries, 78416 Aubergenville Cedex, France

Abstract. By 2025, Battery electric vehicles (BEV) and plug-in hybrid electric vehicles (PHEV) will probably account for more than 20% of the total production of vehicles. With the development of specific vehicle platforms for electric cars, NVH strategies and technologies may strongly change and with them the challenges for NVH engineers.

The integration of the battery pack system in the vehicle is steering the focus towards an "integrated" design of the floor system. Consequently, the design of NVH floor treatment, body in white and battery pack configuration has to be engineered simultaneously to enhance performance and weight reduction. From this, the need for CAE tools which are able to support the integrated design is recognized, even more so in the early stages of development.

The aim of this paper is to analyze how existing and well-established CAE methodologies, such as Statistical Energy Analysis (SEA) and Finite Elements Methods (FEM), can support the NVH design of BEV floor systems. This includes how they can be exploited for the optimization of an integrated NVH package, including not only the carpet and damping pads, but also battery sealing and treatments between the battery and body.

In the first part of the paper, the test setup for the floor insulation assessment of the Jaguar I-Pace, selected as vehicle test case, will be illustrated. Transmission loss measurements in a coupled room and their correlation with FEM simulations are presented thereafter. Simulation will be then applied to explore the possibility of optimizing the positioning of the abovementioned sealing and its impact at vehicle level. Last, the impact of sealing and battery gap treatments on the structure borne noise performance of the floor system are discussed.

Keywords: CAE · SEA · FEM · BEV · Jaguar I-Pace · Battery treatment

W. Siebenpfeiffer (Hrsg.): *Automotive Acoustics Conference 2019*, Proceedings, S. 28–46, 2020. https://doi.org/10.1007/978-3-658-27669-0_3

By 2025, battery electric vehicles (BEV) and plug-in hybrid electric vehicles will probably account for more than 20% of total production of vehicles. Requirements linked to connectivity, autonomous and shared driving, sustainability and electrification are steadily changing the architecture of these vehicles. In the early days of BEVs, original equipment manufacturers (OEMs) mostly carried over technologies used on combustion engine vehicles to electric cars. With the development of specific vehicle platforms for electric cars, noise, vibration & harshness (NVH) strategies and technologies may strongly change and with them the challenges for NVH engineers.

Whether it'll be on a specific BEV platform or shared with an internal combustion engine (ICE) derivative, a floor system with good NVH performance is needed. Electric motor radiated noise is an industry challenge, as well as the noise radiated from the tires, which is becoming louder and louder with the progressive reduction of their rolling resistance. These issues, combined with the ever increasing focus on vehicle weight and its relationship with the battery range mean that even for BEVs, the optimization of the acoustic package of the floor system represents an important challenge for the NVH engineer.

While Jaguar has had much success with launching its first BEV, the I-Pace, continuous development to understand and improve efficiencies is critical to maintaining the brand's competitive position in the market. It's no secret that the mass of the battery can be unlocked for acoustic benefit, however quantifying this benefit is critical to help develop a system approach to the floor which typically includes the design of the carpet, body in white and battery pack. Given the integral nature of the battery and lengthy development and supply lead times, battery definition and integration is frozen early in the programme development cycle. Furthermore, physical testing and prototype validation is time-consuming and costly given the increasing complexity of BEVs. Hence the need for CAE tools which are able to support the integrated design is recognised to help to optimise the NVH performance, cost and weight of the floor system.

The aim of this paper is to analyse how existing and well-established CAE methodologies, such as SEA and FEM, can support the NVH design of BEV floor systems and how they can be exploited for the optimization of integrated NVH package including not only floor carpet and damping pads, but also treatments between the top of the battery pack and the body in white.

NVH physical testing on the floor system of the Jaguar I-Pace and related results are presented in Section 1, while section three provides a numerical/experimental analysis with focus on how the different components of the floor (carpet, vehicle body, battery) contribute to its insulation performance. Section 4 focuses on the exploitation of these data for designing and optimizing the NVH performance of the integrated floor system by means of SEA.

Section 5 closes the paper with some considerations on the structure borne noise performance of the same floor system.

1 Experimental Framework

1.1 Testing Method

In this study, the sound intensity approach is applied for the measurement of Sound Transmission Loss (or Transmission Loss or TL) of the Jaguar I-Pace floor [1].

The TL with the sound intensity approach is calculated through the direct evaluation of the difference between the level of the intensity incident on (Li) and transmitted (Lt) through the test partition (1):

$$\mathrm{TL} = \mathrm{Li} - \mathrm{Lt} \tag{1}$$

with all quantities expressed in decibel (dB).

Since in the emission room the relationship between the intensity level, Li and the averaged Sound Pressure Level Lp, can be expressed [2]:

$$\mathrm{Li} = \mathrm{Lp} - 6 \tag{2}$$

the TL can be derived as:

$$\mathrm{TL} = \mathrm{Lp} - \mathrm{Lt} - 6 \tag{3}$$

Thus allowing the calculation of the TL by only measuring the sound intensity in the reception room and the sound pressure level in the emission room.

1.2 Testing Set-up

Even though the formula (3) is very practical, the effort to build the measurement set up needed for a TL test with the sound intensity method has not to be neglected. The measurement is realized in two reverberant rooms vertically coupled (Fig. 1). To be able to measure the TL of the floor of the vehicle, without destroying it, the car cabin itself has been used as the reception room coupled with a reverberant room as emission room.

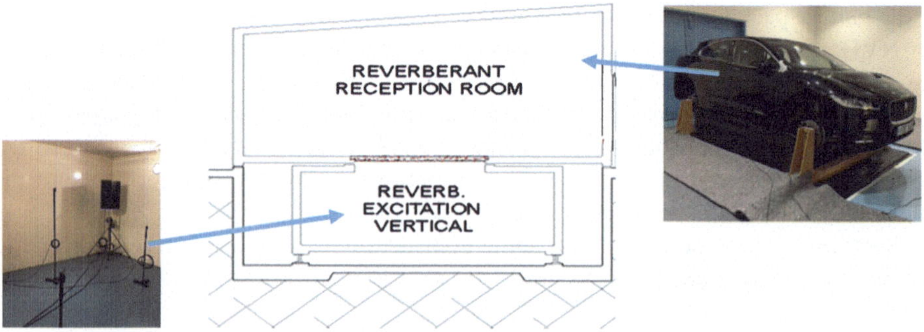

Fig. 1. Sketch of measurement set up.

A system suitable to install and un-install the 605 kg of battery in a convenient way has been built, in which the car is elevated and supported by four metal pillars. Nevertheless, every time it is required to dismount and mount again the battery, an effort of 20 man-hours is needed (Fig. 2).

Fig. 2. Battery disassembly procedure

The measurement is carried out respecting the procedure as below:

a. White noise was generated in the source room by four loudspeakers (two for the low-medium frequencies <2500 Hz and two for the high frequencies, up to 8000 Hz) placed in the opposite corners of the emission room. The noise level is controlled by an amplifier (Crown macro-tech 1202 & 602) and an equalizer (BOOST EQ-231)
b. The acquisition system is B&K LAN-XI module with measurement software Pulse.
c. In the emission room, three microphones (B&K 4189 A21) are used to determinate the average SPL in the source room. They are placed in optimized positions (previously determined) which can offer a representative average SPL of the emission room.
d. In the car cabin, the transmitted sound intensity is measured directly using the B&K Sound Intensity Microphone Probe type 3599 equipped with a remote control to operate the measurement easily.

The microphones with 12 mm spacer were chosen which works well in the frequency range between 400 Hz and 5 kHz, with an accuracy of ± 1 dB. The acquisition time is fixed to 60 s and the distance to the floor is kept to 10 cm. For the measurement of each configuration, at least three repetitions are performed, with the condition that the results are within a tolerance of ± 1 dB. Then the average of these tests is used to calculate the TL.

In order to avoid reverberant field effects on the measurement accuracy, the instrumentation panel in the vehicle is covered with a thick absorbent material (Fig. 3d). Moreover, to avoid leakages the tunnel console has been removed and the surface covered with soft felt material and heavy layer (Fig. 3a). The same treatment has been applied to the hole under the seat covered by acoustic 'Max Pack' (Fig. 3b). In the reception room where the car is positioned, 40 mm of felt was added on the floor to reduce the acoustic level outside the car (Fig. 3c).

To guarantee that the noise level measured in the vehicle is only originated by noise going through the floor of the car and not through the walls or the vehicle sides, the interface between the car floor and the emission room is sealed and the structure is sealed where possible with soft felt material and heavy layer.

The measurement area of the floor panel has been divided in four parts, namely 'driver side', 'front passenger', 'rear passenger' and 'under rear seat' as shown in Fig. 4. This provides more bandwidth for investigation and analysis.

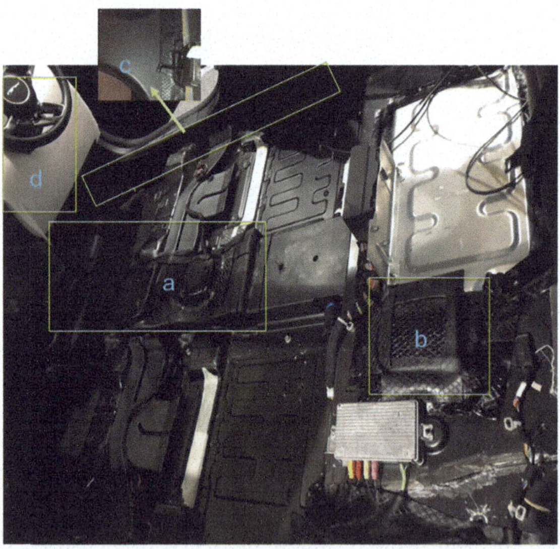

Fig. 3. Details of measurements set up a) console removed + 'Max Pack' covered, b) hole under seat covered by 'Max Pack', c) heavy layer from lower A-Pillar to lower C-Pillar, d) an absorbing layer covers the IP

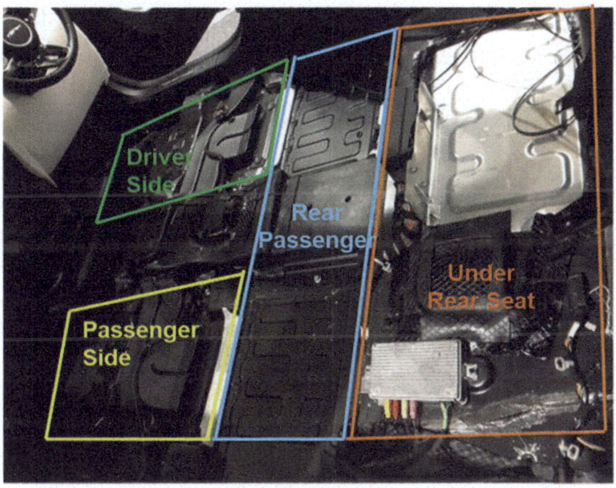

Fig. 4. Measurement area definition

2 Numerical Framework

Car floor acoustics is a topic which has been largely investigated. Engineers can easily differentiate, choose constructions, and further define the specific Bill of Materials (BOM) in terms of density and formulations. Performance specifications are clear and widely accepted, e.g. Insertion Loss (IL) and absorption are common quantities used to set acoustic targets. Additionally, well-established simulation tools allow predicting the pile-up performance at different levels. FEM or the Transfer Matrix Method (TMM) and SEA can support the engineer wherever a refined decision needs to be made [3, 4].

The real challenge today is to verify how resilient these well-established methods are to tackle the complexity of the new electric vehicle architecture. The battery pack integration creates a new small cavity between the floor system and the battery which shall not be neglected.

Although building up a FEM model requires a higher level of effort with respect to the TMM, it also delivers a higher accuracy and the possibility to include high-fidelity details into the problem at hand. That's why, for this analysis an FE model, able to correctly represent the acoustic behaviour of the small gap between the battery and the floor, has been created. After this, evaluations with TMM and SEA have been performed.

2.1 Transmission Loss Simulation

An example of how finite elements can offer a flexible and yet efficient tool to handle vehicle acoustics is given in [4]. In the mid/low-frequency range, advanced numerical

FE schemes are necessary to capture a detailed performance. Additionally, for such a low-mid frequency region, using FE is also a viable solution for predicting the effect of structure borne noise [5].

For this application, the software FFT Actran [6], well-known for its extensive acoustics numerical computation capabilities, has been used. The simulation setup has been thought to represent with high fidelity the measurement setup, even though simplifications have been necessarily introduced, in order to reduce the number of elements and the size of the model and thus to guarantee affordable computational times.

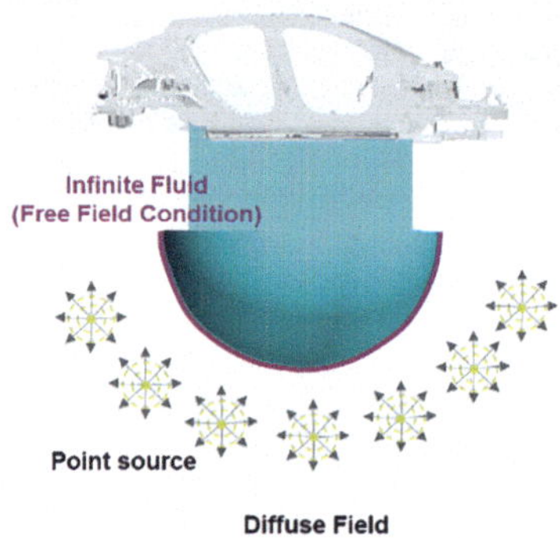

Fig. 5. Simulation set up

As depicted in Fig. 5, the cavity meshed to represent the reverberant room has a 'mushroom shape' so to represent not only the reverberation room but also the presence of the wooden box which connects the car floor to the room. Adaptive meshing has been used to improve simulation performance, at 500 Hz the acoustic mesh counted around 270000 nodes. As for the boundary conditions, to recreate the diffuse field present in a reverberant room, a cloud of point sources surrounding the hemisphere surface was created. These sources are automatically created by the software once their number and distribution are specified. In order to simplify data post-processing, Actran was asked to calculate the incident power, generated by the cloud of sources, near the bottom of the battery. To do so, Actran requires a free-field condition which, in our specific case, was applied to the hemispherical surface. In the model two other cavities have been created, one between the battery and front floor and another between battery and rear floor, so to be able to represent many sealing conditions possibly needed for the analyses presented in the next section (Fig. 6).

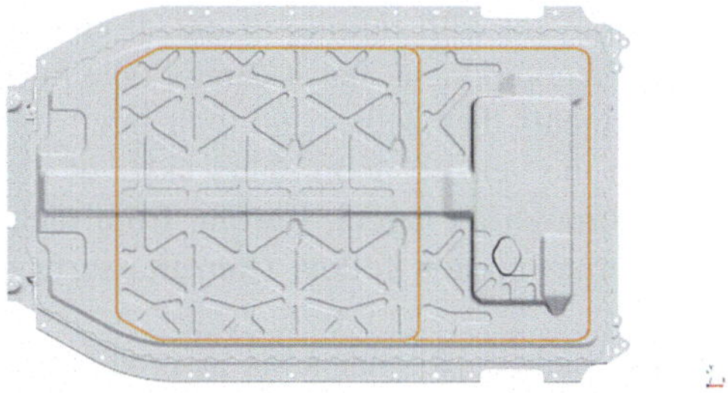

Fig. 6. All possible sealing locations (marked orange)

The vehicle floor has been divided into four areas as in the measurement set up (Fig. 7a). For each of these areas the radiated power has been computed. The incident power on the battery is computed by means of an auxiliary surface meshed below the battery (Fig. 7b). To correctly represent how any modification in the battery gap cavity could impact the overall NVH performance, the response of the full vehicle is not needed, only the floor and the under rear seat are areas of interest. Therefore, in the Actran simulations only the restriction of the modal basis of the full vehicle to these areas was considered. Modes were calculated for the full vehicle with MSc Nastran from 290 Hz to 1035 Hz, totalling 4676.

Such strong simplification is necessary to avoid excessively long computational time but it does not compromise the sensitivity of the results to modifications in the sealing conditions, as the correlation in Sect. 4 will demonstrate.

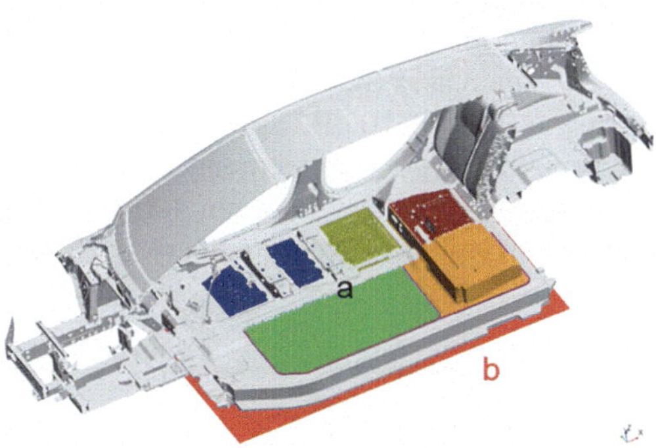

Fig. 7. a) Subdivision of the output as per measurement set up b) battery loading surface

2.2 Airborne Design by Means of SEA and TMM

SEA and TMM are well-known methods used by NVH engineers to simulate respectively the performance at vehicle level and at part level [7].

TMM assumes 2D multilayers with infinite extent. It enables quick computation of the TL and absorption of a component, but it cannot take into account the detailed shape of the acoustical treatments themselves and their interaction with the structure. It relates the pressure p and the velocity v on either sides of a composite by analysing the wave propagation through its layers, starting from their material properties. Reflection, absorption and transmission follow from the surface impedance and the resulting pressures.

SEA is a vibroacoustic method [8] that aims at predicting the energy transfers from one subsystem to another. The coupling power proportionality, the main result of SEA, states that the power exchanged between two subsystems is proportional to the difference of their modal energies, as long as the SEA assumptions are fulfilled [9]. The flexibility in the definition of subsystems, not necessarily as physical components but as a group of modes where the energy is equally distributed and fast computational time makes SEA suitable for many industrial applications [10, 11].

3 Transmission Loss Analysis

In the following paragraph, comparison between measurement and testing results will be illustrated. Firstly, we will analyse the impact of the battery system on the floor system transmission loss, then how adding sealing can influence the performance.

3.1 The Impact of the Battery on the NVH Performance

When the TL of the floor system is measured with and without the battery (Figs. 8 and 9), without any sealing or interior floor treatment, it is interesting to see how below 600 Hz, the battery presence increases the value of the insulation by 4 to 6 dB, while above 600 Hz the increase is around 10 dB. Simulations (in the set up described in Sect. 2.1) and measurements show good agreement, even though simulations generally overestimate the impact of the battery between 400 and 500 Hz. For the sake of conciseness, only a selection of results is presented here.

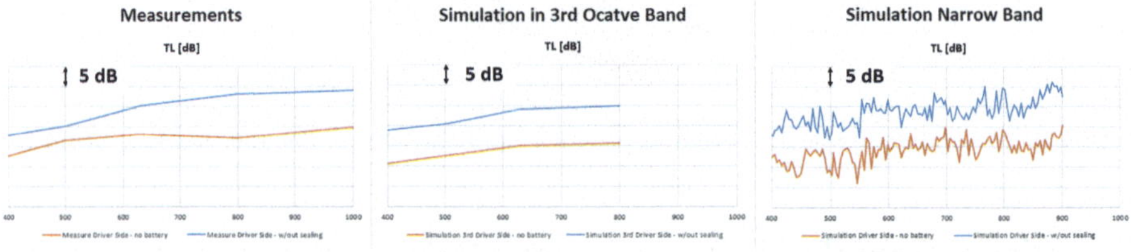

Fig. 8. Floor TL performance with/without battery for 'Driver Side' zone

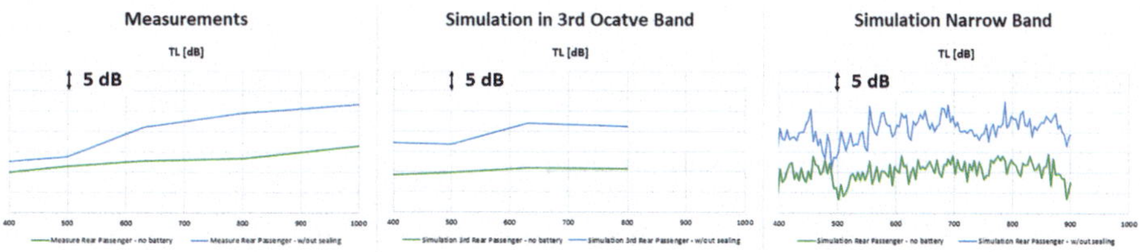

Fig. 9. Floor TL performance with/without battery for 'Rear Passenger' zone

3.1.1 Mass Law Approach

The introduction of the battery has important consequences on the insulation performance of the entire floor system. It is important to clarify, though, that a simple mass-law sound TL will not be enough quantify this change. For low frequencies the sound insulation follows the mass law that can be written for random incident sound waves as;

$$R = 20 * \log\left(\frac{\pi f m}{\rho c}\right) - 5\,(dB) \tag{4}$$

Where f is the frequency, m the area specific mass, ρ the density of air and c the speed of sound in air. Equation (4) tells that for a given structure the sound insulation increases by 6 dB for a doubling of the frequency. In the same manner, for a given frequency the sound insulation increases by 6 dB for a doubling of the mass.

If the formula is applied considering the weight of the battery, it will dramatically overestimate the impact of the battery on the TL of the floor system as simulations with TMM show (Fig. 10). To adjust 'measured' vs 'predicted' TL for the Jaguar

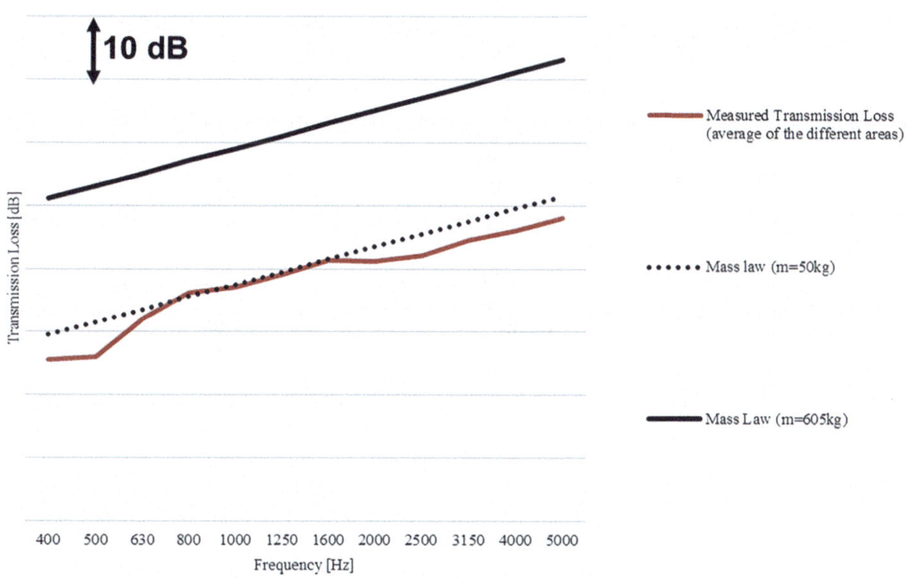

Fig. 10. Battery weight impact on insulation performances.

I-Pace, we should divide the actual battery mass by a factor of 12 when considering only the mass law approach.

3.2 Insulation Enhancement with Sealing

To assess the impact of the sealing and to validate the simulation set up, three different sealing conditions are here examined: no sealing (0%), full perimeter sealing (100%), standard I-Pace sealing which corresponds to ~80% sealing coverage (Fig. 11).

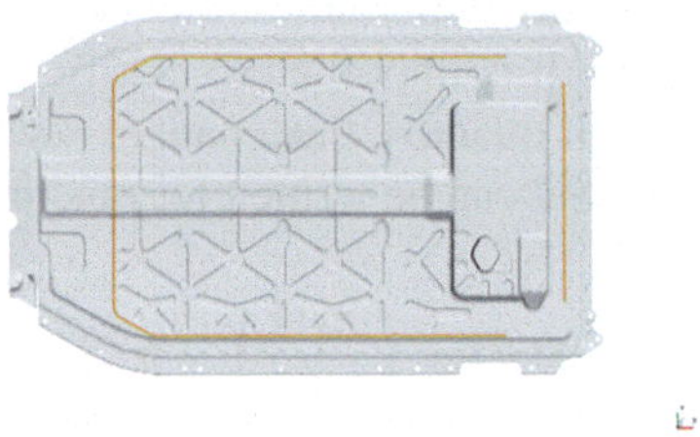

Fig. 11. I-Pace original sealing

Generally, the presence of the sealing increases insulation performance from 1 to 3 dB in the full frequency range (Figs. 12 and 13). The difference in terms of performance between the configuration with original sealing and the one with 100% sealing is small, almost none for some frequency bands of the TL of the front driver area: this is obvious when considering that the main difference between these two configurations is towards the rear end of the battery cover. Such delta becomes more evident for the 'Rear Passenger' area and not negligible (~1 dB) for the 'Under Rear Seat' area (Fig. 14). This means that when designing the treatment, if it is not possible to achieve the 100% sealing condition for manufacturing constraints, it is very important to decide where to place any openings, because they will directly and locally affect the acoustic benefit brought by the sealing.

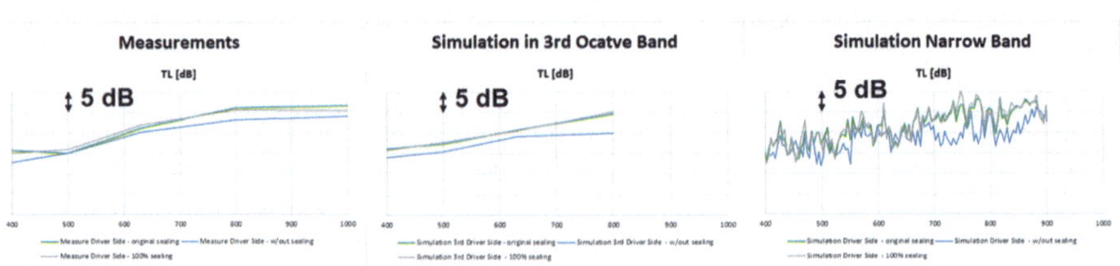

Fig. 12. Floor TL performance with no sealing, with original sealing and with 100% sealing for 'Driver Side' zone

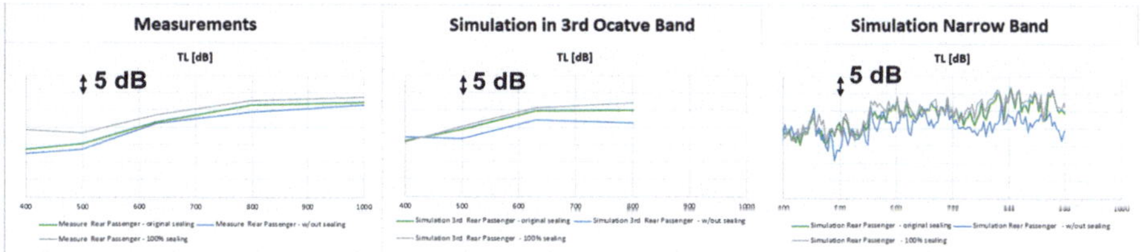

Fig. 13. Floor TL performance with no sealing, with original sealing and with 100% sealing for 'Rear Passenger Side' zone

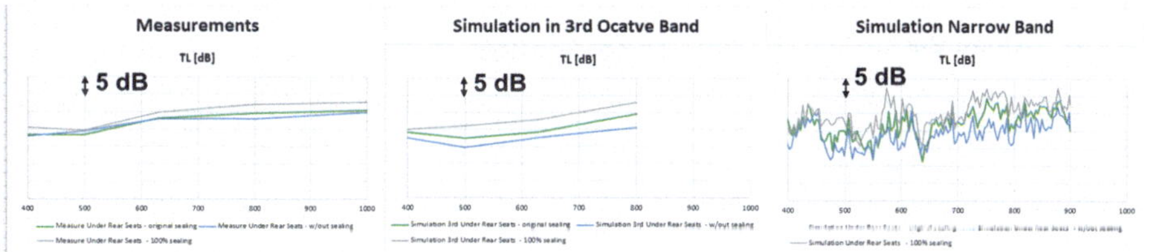

Fig. 14. Floor TL performance with no sealing, with original sealing and with 100% sealing for 'Under Rear Seat' zone

By means of the results presented here it is possible to demonstrate not only that FE simulations are sensitive enough to provide meaningful insight into the impact of sealing solutions but also that they are sensitive to small changes in the system. Therefore, considering the tremendous effort that un-mounting and remounting the battery requires, only simulation will be used to study sealing optimizations.

3.3 Optimization of Sealing Positioning

To study how the airborne noise performance of the floor system can vary in function of the sealing location, the following configurations have been analysed only by means of 3D Actran simulation:

- Open back (Fig. 15a): the further back line of sealing is open;
- Longitudinal sealing (Fig. 15b);
- Transversal sealing (Fig. 15c).

As one could expect, with the open back sealing condition, where the line of sealing closest to the rear tyres is moved in the front of the electrical connection box, there is no relevant consequence of the insulation performance of the front panels (Fig. 16). On the other side, the rear panels, particularly the under rear seats panel, register a visible degradation of TL (~5 dB) especially at high frequency. This result has to be properly taken into account by NVH engineers and designers, especially because many electric vehicles are equipped with an electric motor or range extender in the rear area thus increasing the need of insulation under the rear seats to enhance acoustic comfort.

Considering that NVH engineers and designers are always faced with the challenge to balance different requirements, it is not absurd to think that it will not be possible to seal 100% of the cavity between the battery and the floor. In this case, the guideline should always be to seal in the longitudinal direction and eventually sacrifice the transversal, because the overall performance of the sealing is driven by its longitudinal part (Fig. 17). Indeed, the transversal sealing alone has a very limited impact

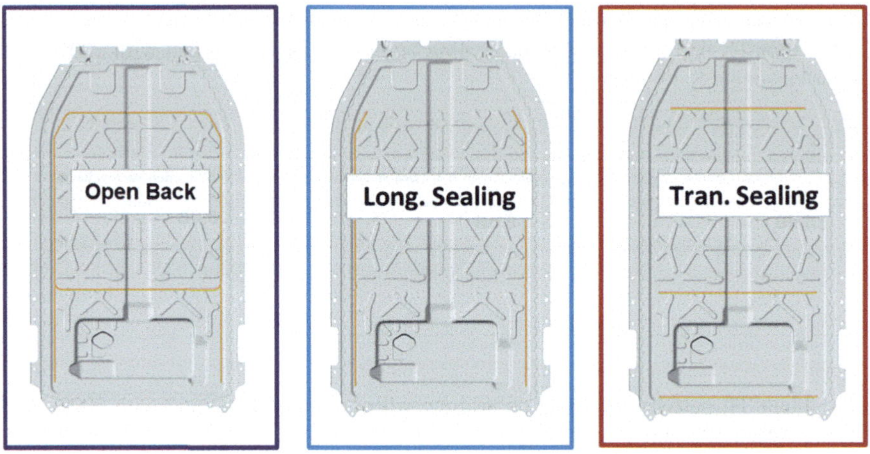

Fig. 15. Sealing configurations: a) Open back, b) Longitudinal sealing, c) Transversal sealing

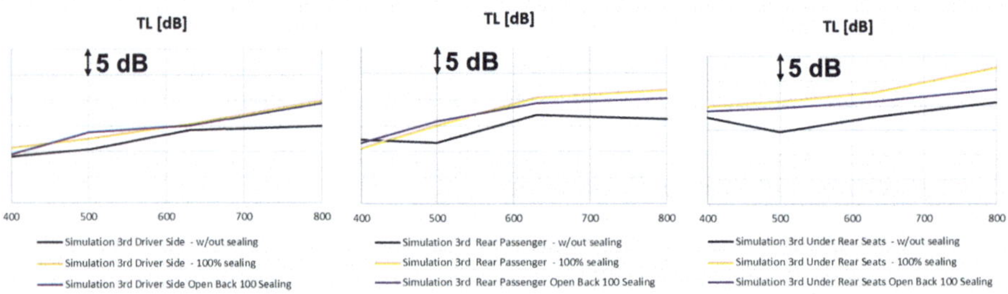

Fig. 16. Impact on TL of different sealing configuration

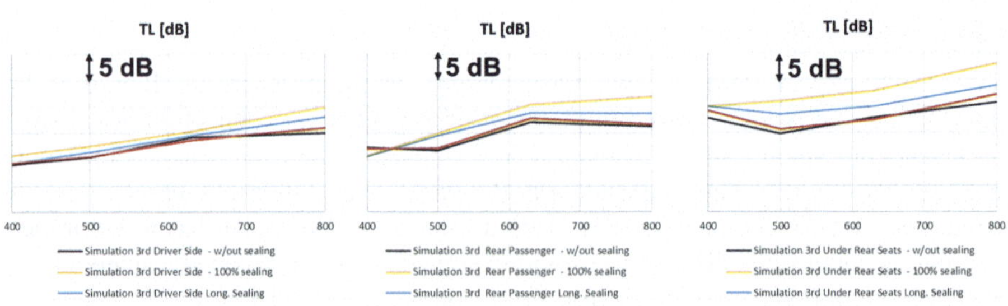

Fig. 17. Impact on TL of longitudinal and transversal sealing

on the overall TL, while in combination with the longitudinal part it guarantees the best performances possible (the difference in terms of TL between longitudinal and 100% sealing configuration is not negligible). Nevertheless, a small gap in the sealing (Fig. 10) needed for water drainage, for cables or other needs, do not create dramatic loss of performances, as the results in Sect. 3.2 are showing (Figs. 12 and 13).

The trends showed by these results seem to be confirmed also by other experimental activities, even though the exact impact in terms of a delta in dB strongly depends on the geometry of the vehicle.

4 Battery Integration in the Airborne Design of the Floor System

4.1 Impact on HV Battery on Vehicle Airborne NVH Performances

To evaluate the impact of different sealing conditions on the overall NVH performance of a vehicle, SEA is applied in order to simulate acoustic transfer functions (ATFs) from the power unit and tyres to the driver and rear passenger head position. Details on the simulation methodology are beyond the scope of this paper and they can be easily found in literature [8, 11].

The SEA model of a vehicle compatible in shape and excitation with the Jaguar I-Pace has been exploited to, case by case, modify the performance of the floor in order to represent the impact of the different battery/sealing configurations at vehicle level. Due to confidentiality reasons, it is not possible to show detailed results but a brief summary is given.

Regarding the ATFs, a volume-velocity source is located in front of the vehicle to represent the noise originated from an electric motor with microphones being positioned at the occupant's ears (Fig. 18). The completion of the sealing to 100% improves the vehicle performances, up to 1 dB, especially in the mid-low frequency. Considering the rear tyres as the noise source, the improvement registered can be up to 1.5 dB.

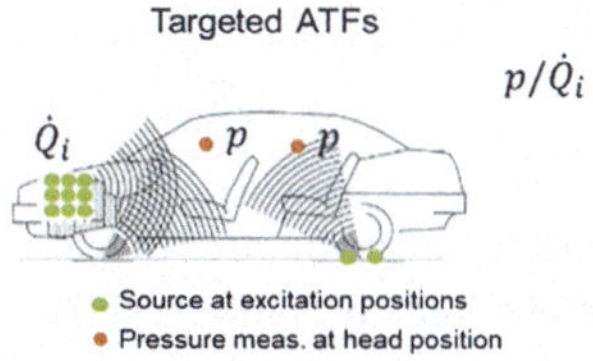

Fig. 18. Schematic representation of the ATFs

The improvements obtained by means of sealing can be exploited in the design phase to optimize the weight of the carpet system. The overall high level of insulation performance of the integrated floor system seems almost to invite NVH engineers not to be afraid of limitations of packaging space and to explore more absorptive solutions for the carpet interior treatment.

4.2 Including a Filling in the Battery Gap

Another variable available for the NVH design engineers is the possibility to include a filler material between the top of the battery pack and the body in white. This solution is very appealing because it will allow system integration with interior NVH package by increasing insulation and acoustic performance without having to increase interior packaging space. Of course, also other requirements as thermal heat up, water drainage, self-ignition and battery installation have to be considered when designing such a solution.

For this analysis, it has been decided for the sake of simplicity to apply the gap filler only for the carpet-covered area, thus no filler below the under rear seats panel. The filler material is a light foam with a density of 24 kg/m^3 (Fig. 19). The thickness of the foam has been selected to avoid strong compressional forces (maximum local compression ~20%).

Fig. 19. Details of the battery gap filler

The acoustic benefits of this solution are proven not only for airborne noise (~2 dB as showed in Fig. 20) but also for structure borne, as it will be explained in more detail in the following section.

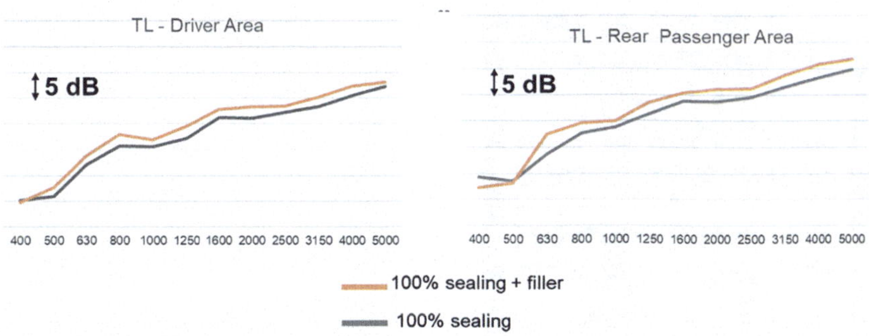

Fig. 20. Battery filler impact on insulation performances.

5 Structure Borne Considerations

Introducing a filler in the battery gap is expected to have an impact also on the structure borne noise performance of the floor system. To realize a fair comparison, the impact of the filler has been analysed with respect to the configuration without treatment and a configuration with an optimized damping treatment. The results here presented are obtained by means of FE simulation with MSc Nastran.

To optimize the positioning of the damping pads, Autoneum has developed a CAE tool, SILVER [12, 13], that processes the mobility result for a given set of load cases and a given set of response nodes. The results of the SILVER are split in two files: the *.aeso file, which contains the ranking of the load-cases, and the *.op2 file, which contains a "colormap" showing the mobility contribution of a given panel and frequency range, as illustrated in Fig. 20. Based on the first output, the engineer can identify for each panel (e.g. floor, rear floor, trunk) which are the most relevant load-cases. Based on the second output, the engineer can identify the most vibrating areas and decide damping pads location accordingly (Fig. 21).

The overall weight of the designed damping package is 1.4 kg, when a standard bitumen damping (density 2 kg/m^3, thickness 2 mm,) is applied. The overall filler weighs 0.75 kg, almost 50% weight saving. Furthermore, it should be emphasized that, usually, the filler is in a compressed state when mounted between the vehicle floor and the battery and this means that both floor and battery cover are in tense conditions. To represent this condition without overly complicating the simulation, a constant tensile pressure load of around 7 kPa was applied to the areas of both components in contact with the foam. This pressure value was set taking into consideration the data coming from a compression test carried out on the foam, a state of compression of the foam of around 30% and the fact that part of the pre-load is carried by the foam. In the simulation, this latter is in a state of traction comparable to the state of real compression in terms of equivalent tension. Therefore, the value here is not the exact pre-load pressure but only a meaningful approximation to be able to see the possible effects that such load can have.

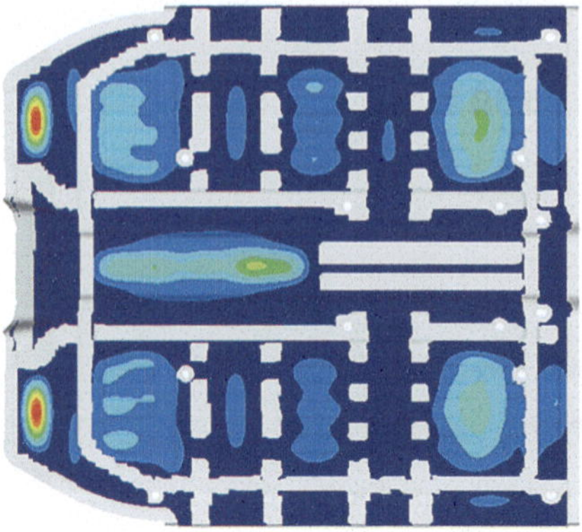

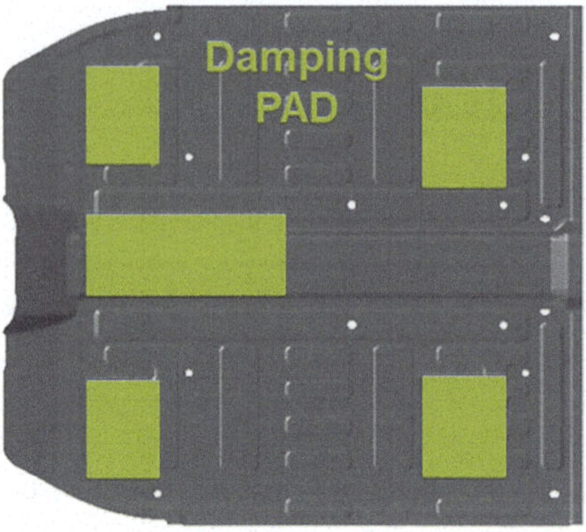

Fig. 21. Battery filler impact on insulation performance.

In terms of performance, the root mean square (RMS) of the acceleration of all the nodes of the floor panels has been computed, when a unit force is applied in the Z-direction where the lower dash panel is welded with the floor panel, as described in Fig. 22. The idea is to exploit the rigidity of the cross beam on the exterior to excite the floor structure in a stiff area. The reason behind this choice of excitation is driven by the measurement set up; even though, the measurement campaign has not been finalized yet, the wish is to measure structural transfer functions while keeping the I-Pace in the same testing set up as described in Sect. 1.2. As a consequence, none of the standard excitation points (for example: sub-frame attachment points) can be used.

In this configuration, the performance of the gap filler is equivalent to the damping performance of the bitumen above 500 Hz and even better between 150 and 400 Hz (Fig. 23). This analysis has been carried out without interior floor treatment and carpet in place.

NVH engineers will be soon addressed by the above as to whether such a gap filler can be a substitute for interior floor damping treatments. The results presented in this section cannot be considered to provide a definite and exhaustive answer to this question, but they bring relevant evidence to the discussion.

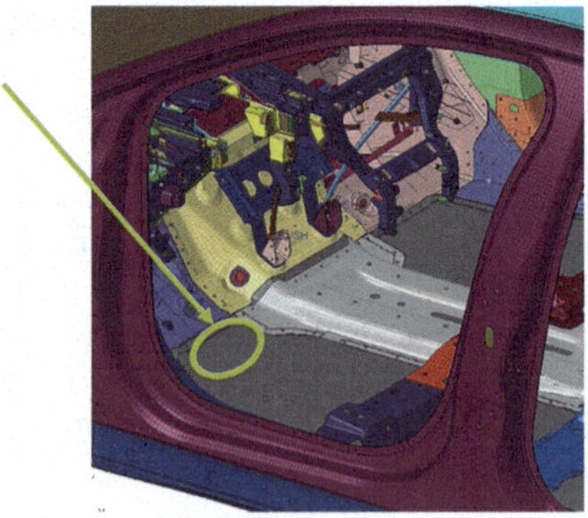

Fig. 22. Excitation: uniform vertical (z-direction) point excitation.

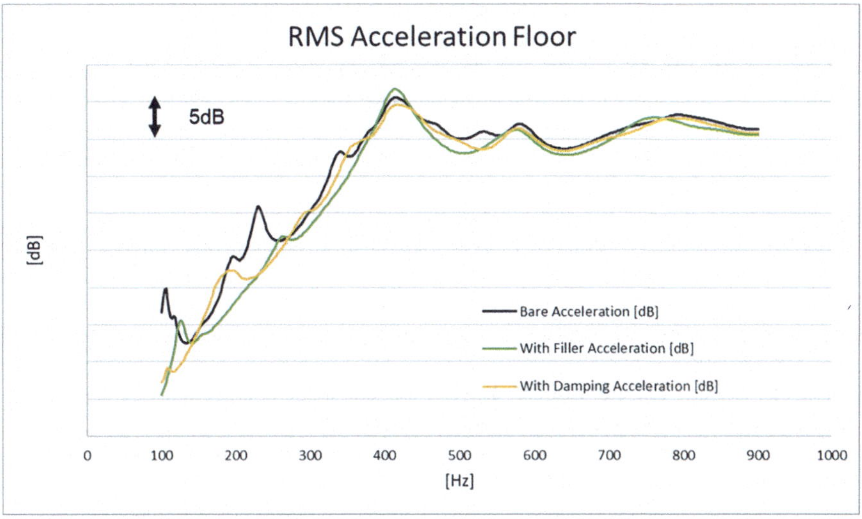

Fig. 23. Battery filler impact on vibrational response of the floor structural panel.

6 Conclusion

From the results presented in this study, it is reasonable to say that the CAE tools available for NVH engineers are ready to face the new, but indeed not so different, challenges that the design of floor systems for BEVs pose. In particular, the integration of a heavy battery below the body floor panels opens the door to sealing and filling treatments, whose study and optimization is required in order to fulfil always more demanding NVH and weight constraints.

From an OEM perspective, the findings of this investigation provide confidence in the ability to accurately predict the benefit of the battery mass in addition to sealing configuration optimisation. It's been shown the level of correlation between predicted and measured performance is good, however a key benefit from the simulation is the ability to rank different configurations in terms of their relative performance. Furthermore, the relative benefits between battery mass and battery sealing have been answered which allows the workload for sealing optimization to be sized and scoped in terms of the relative performance gains to vehicle level performance. Furthermore, physical battery gap filler tests have provided meaningful results which along with the sealing findings will help to shape investigations and ultimately the configurations of JLR products in the future.

References

1. Beranek, L. & al: *Noise and Vibration Control*, INCE publication
2. Crocker, M. G., et al., Measurement of Transmission Loss of Panels by the Direct Determination of Transmitted Acoustic Intensity. Noise Control Engineering, Vol. 17 (1981)

3. Pierce, A. D.: Acoustics – An Introduction to its Physical Principles and Applications Acoustical Society of America (1989)
4. Stelzer R., et al., FE simulation of the Transmission Loss Performance of Vehicle Acoustic Components at Low and Medium Frequencies. SAE International, 2014-01-2081 (2014)
5. Stelzer, R., Delpero, T., D'Amico, R., Assessment of the potential of trim parts to reduce structure-borne noise in the interior of vehicles. In: International Conference on Noise and Vibration, ISMA2018, Leuven (2018)
6. Free Field Technologies, Actran 18.0 User's Guide (2019)
7. Lafont, T., Bertolini, C., Ronzio, F. et al.: Application of Statistical Energy Analysis on a car: from vehicle modelling to parts targeting. In: Proceedings of Internoise, Hamburg (2016)
8. Le Bot, A. Foundation of statistical energy analysis in vibroacoustics, Oxford university press, UK (2014)
9. Lafont, T., Totaro, N., Le Bot, A. Review of statistical energy analysis hypotheses in vibroacoustics, Proc. R. Soc. A, 2014 470, 20130515 (2013)
10. Yang, X.Y. et al., Statistical Energy Analysis of Wind Noise in High-Speed Train Cab, Applied Mechanics and Materials, Vols. 249–250, pp. 307–313 (2013)
11. Keane, A.J. Statistical energy analysis of offshore structures. In: Engineering Structures, Volume 16, Issue 2 (1994)
12. Guj, L., Courtois, T. and Bertolini, C., A FE based procedure for optimal design of damping package, with presence of the insulation trim. In: Proceedings of SAE Noise & Vibration Conference (2011)
13. Yoo, J.W., Lee, S.W., Courtois, T., Horak, J., Ronzio, F. Validation on a vehicle of the structure borne noise of a light-weight body and trim design solution obtained with new integrated FE optimization. In: Proceeding of the Autoneum Acoustics Conference (2015)

A Method for Designing and Delivering Interior and Exterior Noise of an Electric Vehicle

Mark Allman-Ward[1(✉)], Graham Franks[1],
Gianluca Di Nenno[2], and Piercarlo Miglietta[2]

[1] Bruel & Kjaer, Millbrook, UK
Mark.Allman-ward@bksv.com
[2] FCA Italy, Turin, Italy

Abstract. One of the challenges in the development of hybrid and electric vehicles is the interior and exterior noise refinement: for interior noise, the absence of masking by the internal combustion engine during EV operation may be critical due to the appearance of noises that the customer is not used to hear (e-compressor, cooling pumps, electro-hydraulic actuators etc.). On the other hand, the absence of ICE noise may be an opportunity for designing artificial sounds that could be seen as an added value for improving customer's driving experience, perhaps whilst also providing some additional masking of unwanted noise.

For the above reasons, brand requirements and a deep understanding of customers' expectations, may be the most relevant key factors for interior noise development. In addition to this, the new homologation standards for exterior noise leave some design space for a "brand customization" of AVAS.

Although these sounds are not directly connected with the mechanics of the powertrain, nevertheless their link to vehicle's CAN bus parameters (speed, "gas pedal" position etc.) is crucial for a satisfying driving experience.

In this work, an approach has been developed for designing artificial interior noise and AVAS sound for a sporty version of the 500e, where both interior and exterior sounds have been prototyped in an NVH Simulator, allowing nearly all sound prototyping and test work to be accomplished quickly, easily, and efficiently.

The entire process, from the on-track assessment of a "sporty sound" key features, passing through the virtual sound's assessment inside a listening room, to the final delivery of the demo vehicle for jury testing and type approval measurements, is here described.

This approach has then been extended to include the use of panel transducers (PT) to deliver the AVAS sound. PT have the advantage that they are much smaller to package and can even deliver lower frequencies than the small loudspeakers typically used for standalone AVAS sounders. A method for using PT to deliver AVAS is also presented.

Keywords: Electric vehicles · Interior sound · AVAS

W. Siebenpfeiffer (Hrsg.): *Automotive Acoustics Conference 2019,* Proceedings, S. 47–63, 2020. https://doi.org/10.1007/978-3-658-27669-0_4

1 Brand Sound Requirements for a Small, Sporty Electric Vehicle

1.1 Foreword

It is generally known that the NVH development of electric vehicles (EV) is bringing new challenges, mainly due to the missing contribution/masking effect of the internal combustion engine (ICE). Road noise and wind noise are becoming more relevant, and tonal noises generated by the e-motors and other electric actuators and devices are more perceivable and annoying.

Another challenge of increasing importance is differentiation. If we consider sports derivatives of most vehicle lines, one of the key factors in the sporty image of the vehicle is the sound, and almost all the sporty character of the sound has so far come from the ICE. However, increased performance motors for sporty EVs generally have the same sound character as the standard motors, perhaps with the motor whines even more prominent. Furthermore, the increase in the number and diversity of EVs in development (and soon for release) is driving the requirement to differentiate vehicles from the competition, by improving NVH towards the levels seen in ICE vehicles, and further driving the desire for new features and opportunities to differentiate one EV from another.

One of these opportunities is the possibility of creating an interior soundscape that, especially in sport cars, could be an interesting feature from a customer point of view.

In this work, a system that allows the generation and demonstration of an interior soundscape, user selectable between different types, has been investigated for a sport version of a small electric vehicle like the Fiat 500e. The system is also capable of delivering an Acoustic Vehicle Alert System (AVAS) sound which was in this case reproduced by means of panel transducers (PT).

The scope of the project was for the interior sound to include some future-oriented sounds, plus some other sounds inspired by the current 500 Abarth. A selection of AVAS sounds were also required that complimented the interior sounds.

Two separate jury tests were used to select the best solutions developed, one for interior noise and one for AVAS sounds.

2 An Approach for Designing an Interior Soundscape for an EV

2.1 Outline Approach

Essentially, the approach used to design the soundscape for this EV is a process encompassing a series of steps. Initially, it was necessary to define the goal to achieve. This can be done through describing desired or experienced emotions, or as in this project, experiencing existing reference vehicles (or other objects). This goal needs to be communicated as clearly as possible to the whole team involved in the project.

Secondly, an efficient way of working was adopted to allow evaluations and feedback at regular points, by a team that was distributed geographically.

Thirdly, the delivery system needed to be defined in conjunction with the goal and the sound, such that the final sounds could be delivered successfully.

Finally, the sounds, delivery system and vehicle needed to be integrated.

2.2 Defining the Goal and Requirements

In order to explore which kind of sound could be the best option for this electric vehicle, a driving session to compare an existing Fiat 500e with two different versions of the 500 Abarth was organised in a proving ground. The two Abarth vehicles differed in the types of exhausts, one being fitted with an extremely sporty exhaust, while the other was equipped with a "quieter", standard system.

The driving session was attended by the whole team involved in the project, including the sound designer, B&K project engineers, representatives of FCA NVH department and representatives of Fiat Product and Brand Teams.

Driving the vehicles on the track was helpful for the sound designer and the Team to better understand and get aligned with the brand requirements for the sounds to be implemented in the new vehicle, their relationship to vehicle parameters, and response to driving style.

After the driving session, a debriefing was held in which it was decided that the sound of the sporty exhaust was considered the "image vehicle" for the design of the new sounds.

2.3 Creating the Design Environment and Tools

An efficient way of working was needed, where the team could share sounds and driving experiences despite being located in three different countries. Due to the requirement of the sounds to change according to some mix of vehicle parameters, and the vehicle's existing (mostly masking) sounds also changing with vehicle parameters, it was essential to be able to evaluate sound concepts and designs quickly and easily in a driving environment. It was therefore decided to do most of the development of sounds using a desktop based NVH Simulator to provide the vehicle driving sounds and parameters. Apart from the efficiency of being able to work in the sound design studio and drive the target vehicle instantly to check the latest creation, it avoided the requirement for a lot of time spent working in a prototype car which was anyway unavailable. Working in a simulated environment further enabled the simulator model, with the sound designs, to be shared between locations to get feedback from the team.

It was decided that the existing 500 Abarth was a reasonable representation of the masking sounds that could be expected from the concept sporty EV. A quick set of NVH measurements of the 500 Abarth with the target sound direction was made, limited to driver seat interior noise data, and an NVH Simulator model was prepared. The model separated the vehicle masking sounds (the combination of Road, tyre and wind) from the powertrain sounds, meaning that whilst driving in the simulator, the masking sounds alone could be played to represent the masking sounds of the target EV. The powertrain of the image vehicle was left in the model, meaning that at a button click the sound design could be compared with the image vehicle if desired. It was decided not to add the sounds of the EV powertrain to the simulator model at this point, since their level was expected to be low relative to the image sound experience (Fig. 1).

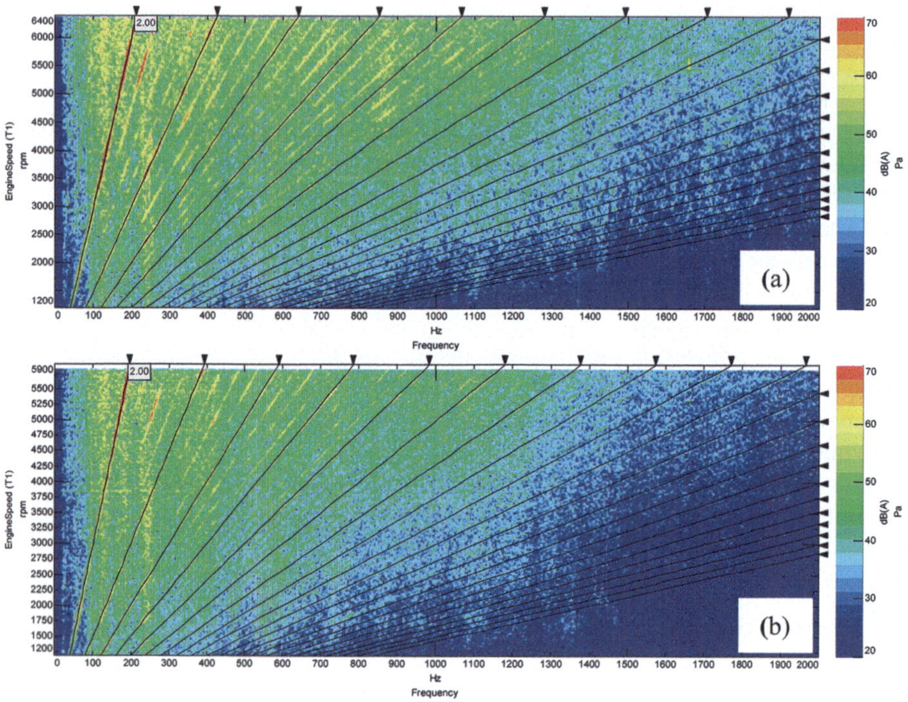

Fig. 1. (**a**, **b**) Interior noise during a fast run-up maneuver (a) and slow run-up maneuver (b) of a 500 Abarth taken as a starting point for the virtual sounds to be implemented into the 500e.

An initial set of progressing sounds were designed, and loaded into the NVH Simulator as a blended wave table using speed and "Go pedal" ("Gas pedal" is not really appropriate for EV) as parameters.

It was quickly realized that although the NVH Simulator is capable of producing conventional vehicle sounds in a fully calibrated and accurate way using many different classes of sound object, the consequence of removing the ICE is that the "rules" of sound progression and harmonic patterns present in conventional vehicles have also been removed. This allows the opportunity and freedom to design sounds that don't have to follow a conventional vehicle sound format, but it also requires new software algorithms to combine and progress sounds according to the design intent.

In order to create a rapid prototyping environment for the sound design, it was decided to use the NVH Simulator to provide the driving experience, parameters and masking sounds, and to use very flexible sound design tools in parallel. Since the system was eventually going to be used in-vehicle, CAN bus was selected as the protocol to transfer vehicle parameters from the NVH Simulator to the sound design system. In this way, the simulator took the role of simulating the baseline vehicle, and the sound design system could be plugged into the simulator or the test vehicle in a totally interchangeable way, with only one button-push to configure CAN bus parameters for either set-up.

The detailed sound design process then ensued, using the NVH Simulator model as the background sound (occasionally with the image vehicle powertrain as a "touchstone"), and a series of different sounds for the interior driving sound were created. Several new sound designs required new source types and algorithms to combine and progress the sound with vehicle parameters. The final sounds were then created from layers of these different sub-components, each one progressing individually to create a continually evolving soundscape.

2.4 AVAS Sound Design

In some ways, AVAS sounds are more complicated to design than interior sounds, because the sounds vary not only with vehicle parameters, but also are perceived by observers including doppler shift, amplitude panning and angle of incidence effects. This means that the sound that was designed in the Digital Audio Workstation (DAW) can sound quite different in reality.

The NVH Simulator again allows an efficient approach. By using the Exterior Sound Simulator (ESS) module, the designed sound can be assigned to a sound source located on the moving vehicle. The simulation of the AVAS sound on that vehicle driving past an observer then includes all these effects. Furthermore, the ESS can apply a transfer function to incorporate the performance of the delivery system. This is significantly quicker than downloading the sound onto a real vehicle and making an assessment in the real world.

A series of 8 base sounds, some with 3 variants of each sound, were created using this technique. 18 sounds were considered good enough to be distributed to the wider project team, where after an initial assessment using the ESS, 8 sounds were put forward for formal jury assessment using ESS in the FCA NVH Listening room. A ranking of sound preference was recorded, and is described later.

The next step was to set-up the assessment vehicle with a system to deliver the designed sounds, both interior and exterior.

3 The On-Vehicle System Installation

3.1 Optimisation of Sound Source Location

A laboratory based study was performed to investigate the optimal locations for sources to deliver the sound designs. 2 possibilites were considered;

- Panel transducers to deliver interior sound and exterior (AVAS) sound
- Panel transducers to deliver exterior AVAS sound only, along with loudspeaker based delivery of interior sounds

3.2 AVAS Source Location Optimisation

Initial testing was performed by measuring vibro-acoustic Noise Transfer Functions (NTFs) by hammer excitation at the circled locations in the pictures below (Fig. 2).

Fig. 2. Investigated panel transducer locations – Front of vehicle.

The chosen locations reflect the requirement for the transducers to be adhesively mounted to the inner surface of a reasonably (but not necessarily perfectly) flat panel, which somewhat limits the available choices.

NTF Measurements were made in the B&K hemi-anechoic chamber measuring Sound Pressure Level (SPL) response at interior and exterior microphones (representing AVAS test locations) as indicated by the red dots in Fig. 3.

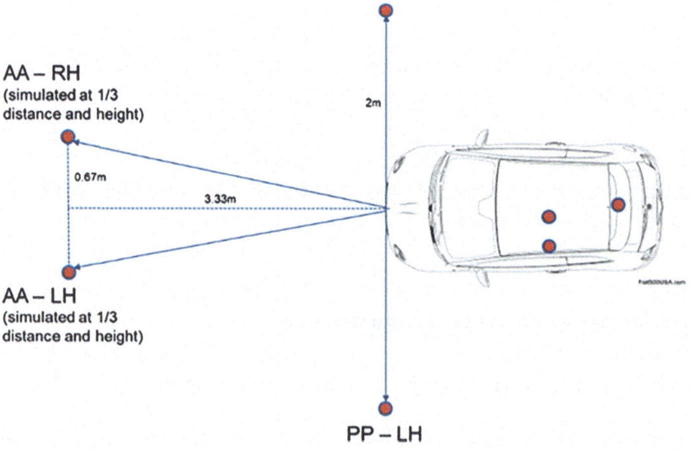

Fig. 3. NTF test set-up – Front of vehicle.

As with the use of a traditional speaker for AVAS delivery, one needs to be mindful of maintaining the optimum delta between high SPL at the exterior AVAS microphone locations and minimal SPL at the interior microphones (assuming that it is desired for the AVAS sound to be inaudible inside the vehicle). Location 2 (LH side of the front bumper) was found to offer the best compromise in this regard, the NTFs for which are shown below in Fig. 4.

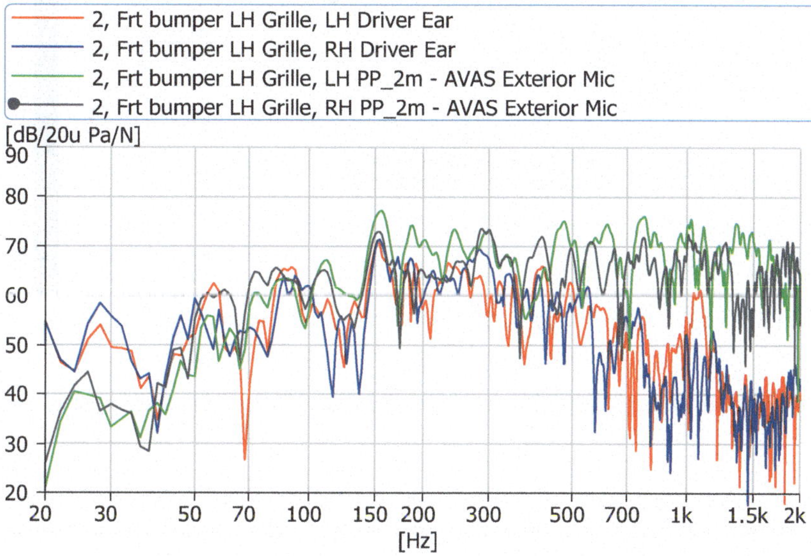

Fig. 4. Interior vs exterior NTF comparison – Location 2.

A simulated AVAS test was performed on the Millbrook ISO Pass-by Track using a broadband signal driving a transducer at location 2 to assess whether a single transducer was capable of meeting the AVAS frequency and level requirements. The results shown below (Fig. 5), indicate that the minimum requirements can be satisfied with this configuration, although the transducer was upgraded to a larger version, providing more headroom, for the final demonstration.

3.3 Interior ESE Delivery Hardware

The laboratory investigations showed that a relatively large number of panel transducers would be required to achieve 'Abarth' levels of very low frequency interior noise, so this option was dropped in favour of a rear mounted Bluacs Technologies 'fresh air' speaker. This device delivers higher levels of low frequency noise from a smaller package when compared to conventional subwoofers and achieves this by installation on the cabin boundary, with the speaker 'porting' through a cabin boundary opening.

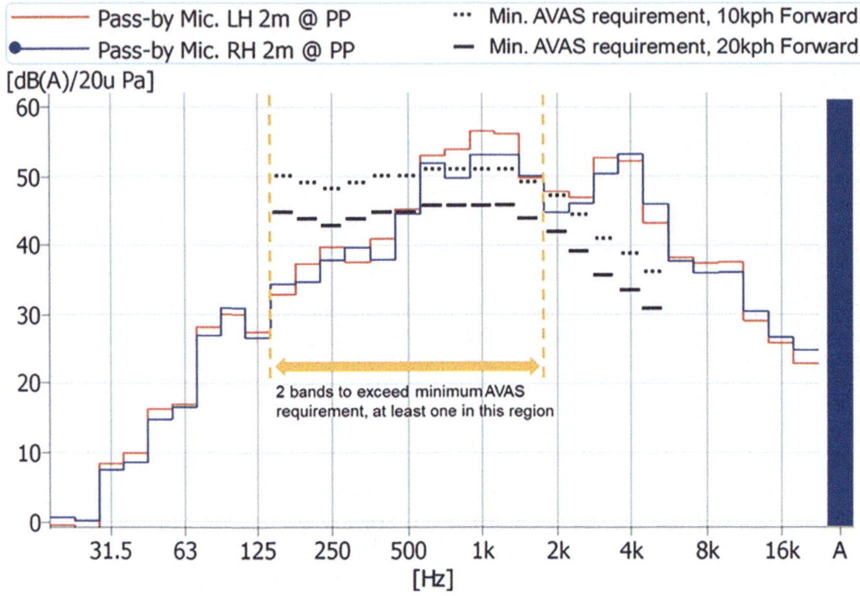

Fig. 5. Simulated pass-by, panel transducer location 2, 7 V pink noise signal.

Figure 6 shows a comparison of the SPL per unit drive voltage for the 'Fresh Air' speaker vs equivalent measurements with a pair of panel transducers on the boot floor.

Since it wasn't permitted to modify the sheet metal work of the test vehicle, we fulfilled the 'Fresh Air' speaker installation requirement by installing the speaker port in place of the existing cabin air extraction vents (Fig. 7) in the rear of the boot floor. The resultant installation offered excellent low frequency sound delivery which also benefitted from being located in the region of the exhaust orifice in the Abarth vehicle. This allowed for a subtle subjective 'connection' with the acoustic environment in the ICE engine vehicle.

The distribution of the sound field in the vehicle is very important to the 'realism' of the driving experience; one needs to have the impression of being enveloped in a 'sound cloud' rather than being able to perceive discrete sources, especially where those are identifiable as loudspeakers.

In order to provide a balanced sound field, the remainder of the interior contribution was provided by a pair of upgraded co-axial door speakers. If time had permitted this could have been improved by extending the hardware further to include high level tweeters or panel transducers in the front upper cabin area.

The hardware configuration used for the demonstration and jury evaluation is shown in Fig. 8 below.

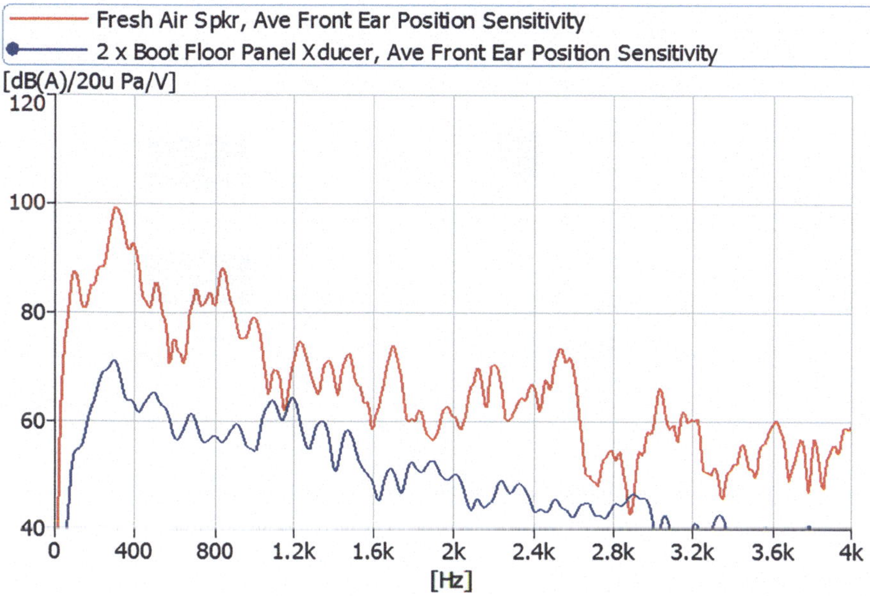

Fig. 6. SPL/Unit drive voltage, fresh air speaker vs twin panel transducer.

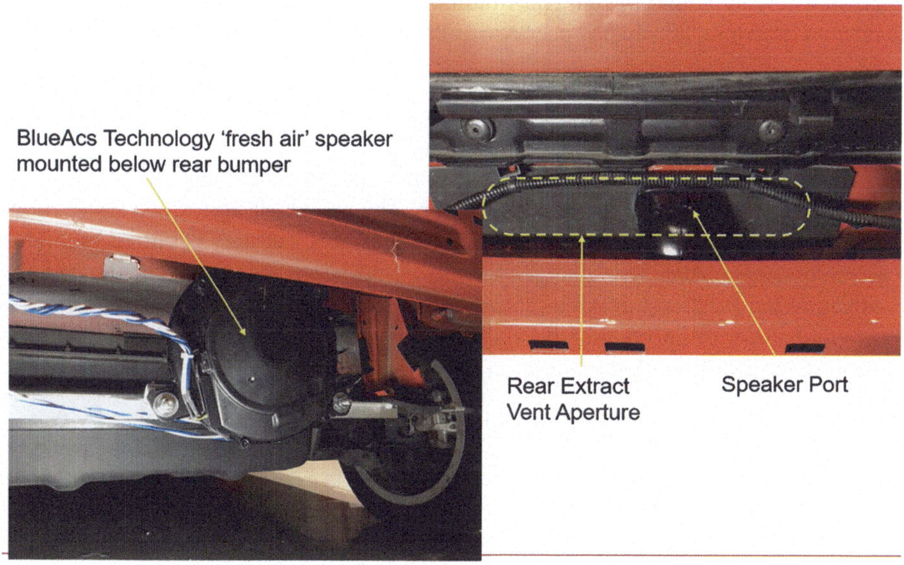

Fig. 7. Fresh air speaker mounting location.

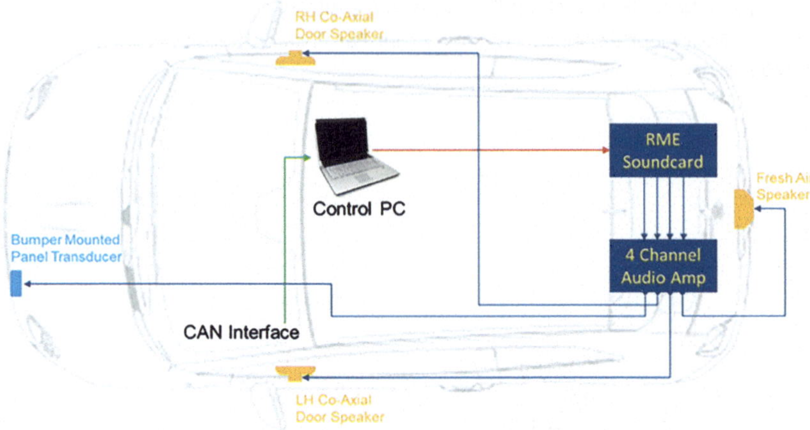

Fig. 8. Schematics of ESE and AVAS delivery hardware.

4 Integration

The final stage of designing the sounds was to insert the designs from the lab into the vehicle. As stated before, this involved plugging the sound design laptop in the vehicle, instead of the NVH Simulator. The results of this were at once both positive and negative. The core sound design from the laboratory development was fairly close to the desired outcome, however it needed adjusting due to several causes:

- The masking noise in the 500e test vehicle was not the same as the Abarth based NVH Simulator model. This created some changes in the low-mid frequency ranges which affected the balance of the sound design.
- The power meter on the dash of the test vehicle had specific bands linked to torque request and therefore "go pedal" depression. The sounds designed had some significant changes throughout the "go pedal" progression, but they occurred at slightly different stages than the changes to the coloured bands on the meter. Simply adjusting the progression points to align with the meter had a significant effect on the impression of an integrated system, perceived quality and subjective rating of the system. It follows that integration of the entire user experience is paramount.

Following Integration testing and tweaking, the test vehicle was returned to FCA.

5 Jury Testing

In the months following the vehicle delivery to FCA, several sessions of jury evaluations were performed at FCA facilities.

Representatives from different departments of FCA were included in these sessions, from Product Development to Product Planning, Quality and Brand (Fiat, Abarth and Mopar).

5.1 AVAS

As stated in Sect. 2.4, the development team shared 18 sounds with the wider project Team. In order to reduce the number of sounds to be evaluated (the jury testing of 20 sounds would have proven to be too long and confusing) a preliminary assessment was performed by a restricted team, in order to get to a reduced number of 8 sounds to be tested by formal Jury procedure.

A jury assessment in the FCA Listening room, using the ESS module of the NVH Simulator, was carried out on those 8 sounds, and a full ranking of preference for the sounds was gathered. Two sounds were selected to be put forwards for real vehicle trials.

5.2 Interior

The vehicle was supplied with 8 interior sounds for evaluation. Several sessions of in-vehicle assessment were run by FCA. In Table 1 the results of the subjective evaluations are reported. Every driver had to express a single choice for one of the 8 proposed virtual sounds, or the basic noise of the electric vehicle when the additional sounds were switched off.

Table 1. Synthesis of the jury evaluation for interior sounds.

Sounds N°	Product development	Product planing	Quality	Brand	Total
Baseline	2	1	0	1	4
2	14	3	3	5	25
3	1	0	1	1	3
4	1	0	1	0	2
5	2	1	1	2	6
6	0	1	0	1	2
7	3	1	1	2	7
8	1	0	0	0	1
9	0	0	1	0	1

Sound n° 2 was the preferred by far, chosen by almost 50% of the people involved.

It is interesting to observe that – even when the aim of the jury test is to evaluate which additional sound is the most appropriate for this particular sporty version of a 500e – the basic interior noise of the electric vehicle with no additional sound was selected as the best solution by around 8% of the drivers. This fact confirms that the absence of the combustion engine sound is a characteristic of the electric vehicles that currently has a strong appeal to customers.

6 Experimental Validation

Two sessions of experimental measurements were conducted on the vehicle, taking into account both interior noise and AVAS sound.

6.1 Interior Noise Measurements

For interior noise, the 8 different samples from the set of designed sounds were measured in the proving ground, together with the standard 500e interior noise (the "Baseline" of Table 1). Both fast and slow run-up maneuvers were measured, and comparisons between the more appreciated sounds are presented in Figs. 9 and 10.

Figure 9a shows the standard interior noise from the 500e, with the main e-motor orders highlighted in yellow. The diagrams of Figs. 9b and 10a,b show the main features of the artificial sound that were added to the vehicle's interior noise (highlighted in red). It is also interesting to note that whilst some of the characteristics in these plots look like "engine harmonics", they don't necessarily sound as such. It follows that we need to take care with interpreting colourmaps of these synthesised sounds.

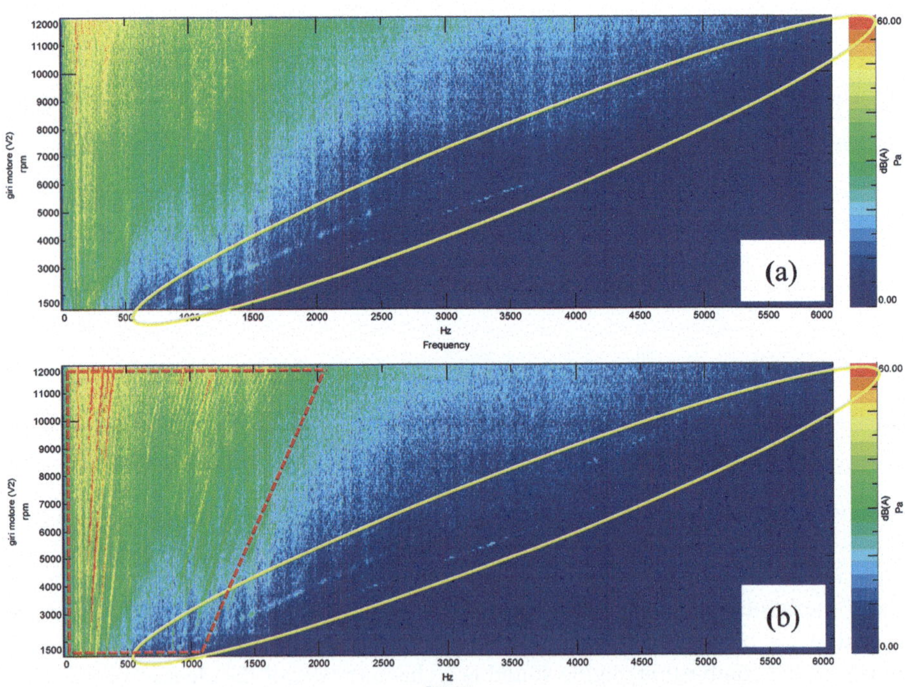

Fig. 9. (**a**, **b**). Interior noise during a slow run-up maneuver for 500e "baseline" (a) and with the artificial sound that was the most appreciated during the jury testing (b).

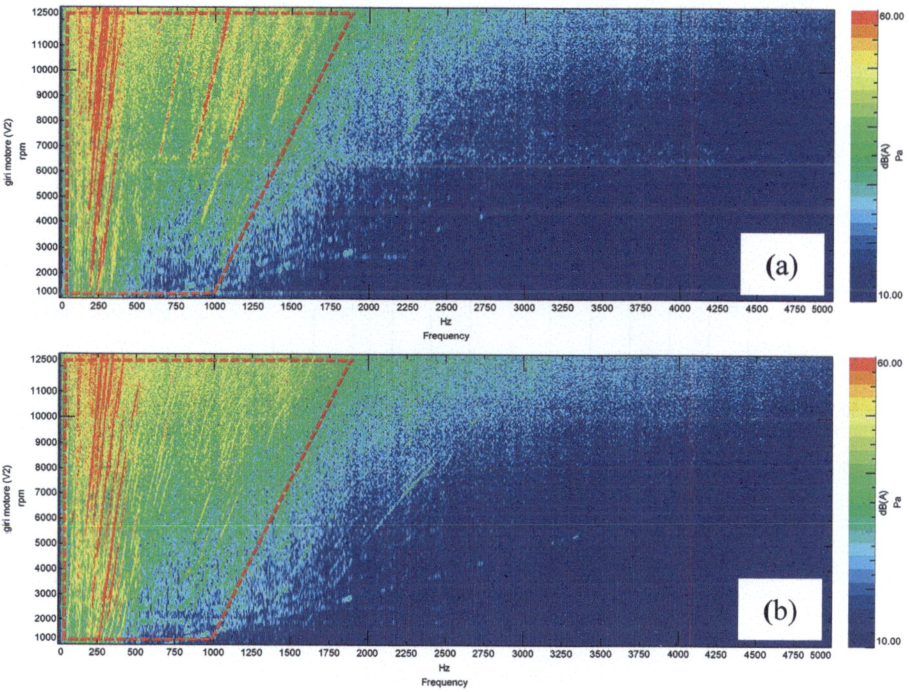

Fig. 10. (**a**, **b**). Interior noise during a fast run-up maneuver ("wide open throttle") for 500e with two different additional sounds that were evaluated during the jury testing.

7 AVAS Type Approval Measurements

As described in Sect. 5.1, the selection of the AVAS sound was done in two steps: a preliminary cull followed by a jury evaluation in the FCA listening room, used to select the two solutions that were considered the most appropriate to be tested in the vehicle. This was then followed by a subjective evaluation in the proving ground, where the final decision on the AVAS sound was taken. At the proving ground assessment, it was possible also to download all 8 sounds to VSound, an on-vehicle DSP powered sound prototyping system, and subjectively rank all the 8 sounds in a real world exercise to mirror the ESS evaluation made in the listening room. The rankings were identical.

In Fig. 11a,b and 12a,b the type approval measurements of the two most preferred sounds are presented. Both of them are compliant to European Regulation [1] limits (highlighted in pink).

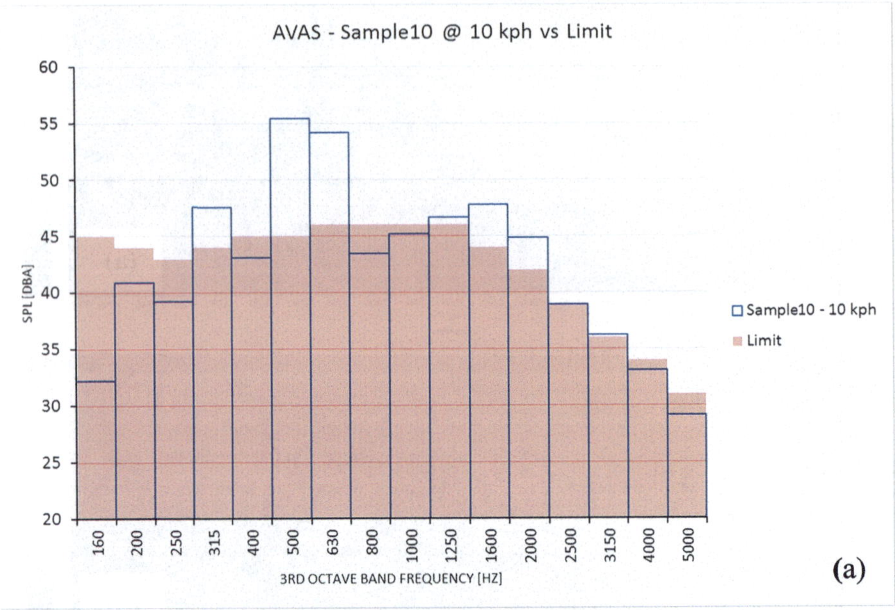

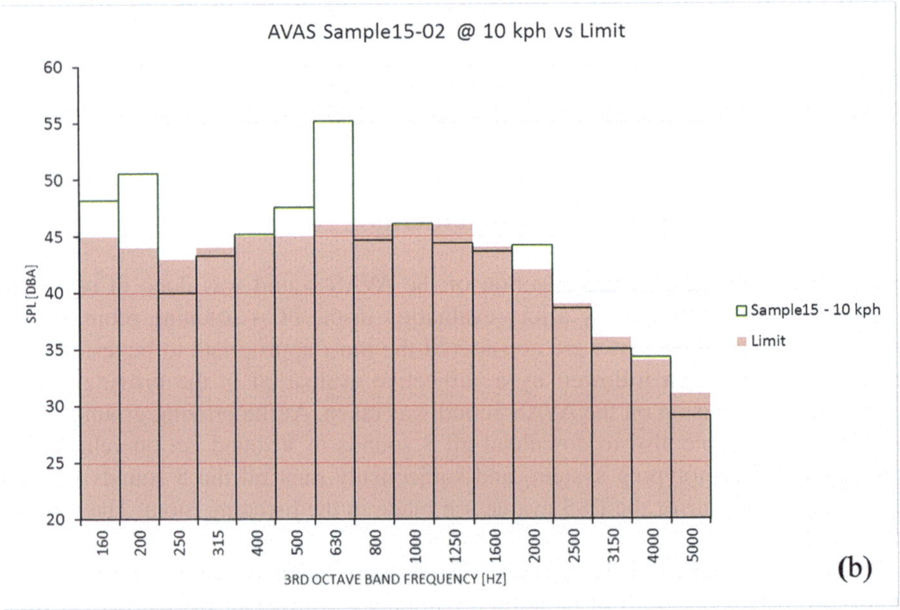

Fig. 11. (**a**, **b**) Type approval measurements of AVAS noise for the two different sounds that were evaluated during the jury testing at the proving ground (speed 10 kph).

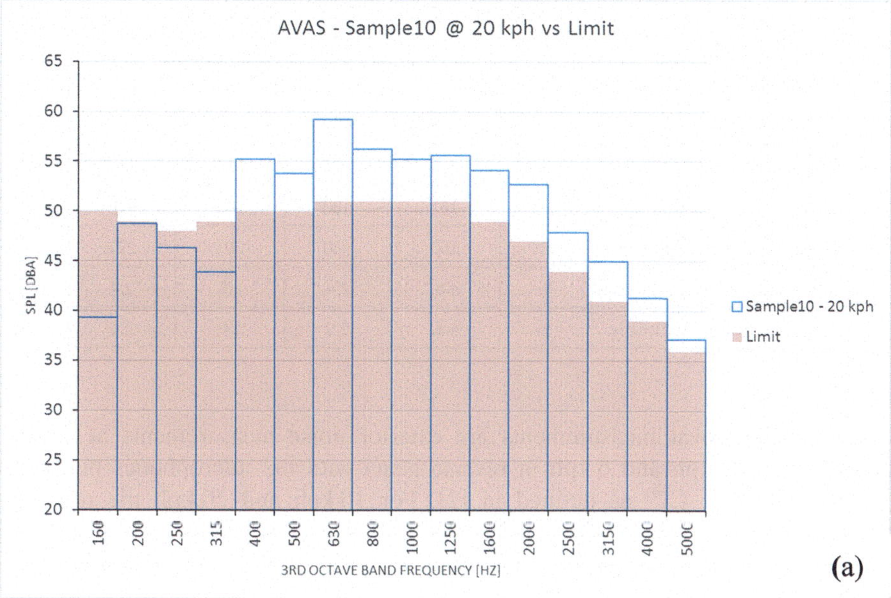

(a)

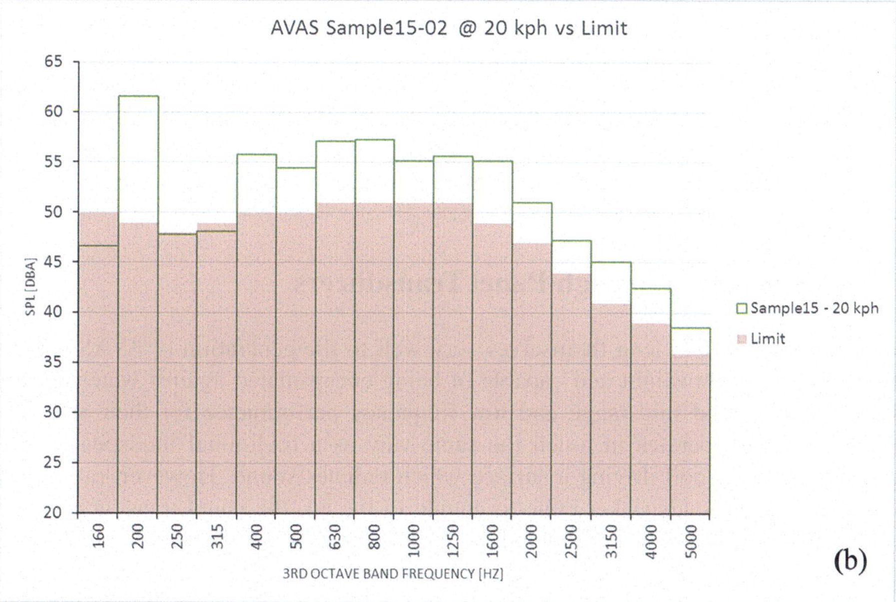

(b)

Fig. 12. (**a**, **b**). Type approval measurements of AVAS noise for the two different sounds that were evaluated during the jury testing at the proving ground (speed 20 kph).

Table 2. Frequency shift measurements for "AVAS10" sound

Frequency shift check		Test Results at Target Speeds			
		5 km/h (Reference)	10 km/h	15 km/h	20 km/h
Reported Speed	km/h	5,1	10,0	14,9	20,1
Frequency, f_{speed}, Left Side	Hz	161	180	201	220
Frequency, f_{speed}, Right Side	Hz	160	180	201	221
Frequency Shift, Left Side	%	n.a.	2,4	2,5	2,4
Frequency Shift, Right Side	%	n.a.	2,5	2,6	2,5

The type approval measurements are exterior noise measurements at constant speed (10 kph, 20 kph and 6 kph in reverse gear) with the microphones put at 2 m from the centreline CC′ as defined in [2]. For 10 kph and 20 kph the regulation requires that at least two 3rd octave bands shall exceed the limits, and at least one of the two bands shall be equal or lower than the 1600 Hz band. For 6 kph in reverse gear, the regulation requires that the measured sound level is above 47 dB(A).

In addition, the requirement that at least one tone in the frequency range 160–5000 Hz increases its frequency at least of the 0.8% per km/h within the 5–20 km/h range was also met, as shown in Table 2.

After the subjective evaluation at the proving ground and the type approval measurements, the selected sound was the so-called "AVAS Sample10". Moreover, the experimental measurements showed the opportunity to reduce the noise level at least in some 3rd octave bands, whilst still remaining within homologation limits.

8 AVAS Sounds Through Panel Transducers

Panel transducers (PT) lend themselves very well to the generation of AVAS sounds, being compact, lightweight and capable of being encapsulated against water ingress, whilst offering good mid-range and low frequency performance for their size. The panel transducer operates in much the same way as a traditional loudspeaker with an electromagnetic coil driving a surface which radiates sound. However, rather than using a dedicated loudspeaker cone to couple to the air, the panel transducer makes use of whatever surface it is mounted to. A schematic of the transducer is shown in Fig. 13.

There are several attractions of the PT as a delivery mechanism. Whilst it has no onboard amplifier, most vehicles already have amplifiers for the infotainment system where CAN bus is already connected, and an extra channel to drive the PT could be easily incorporated. A node on the vehicle can network can therefore be deleted. Furthermore, the PT has a volume roughly equivalent to a hockey puck, which can be

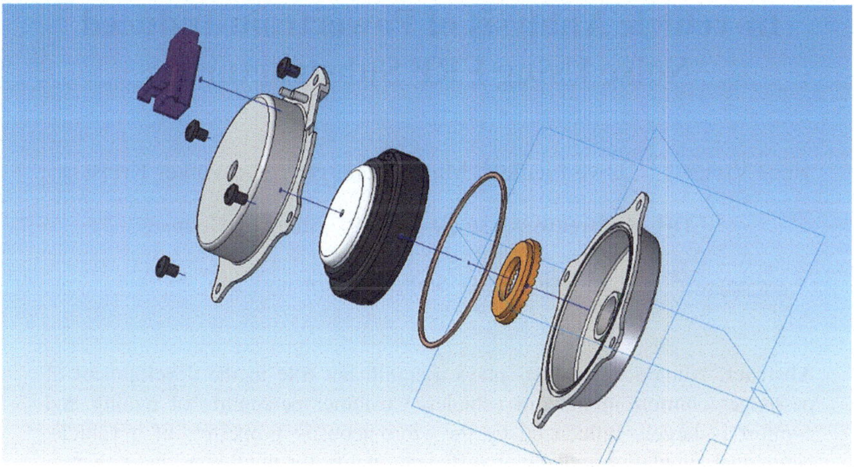

Fig. 13. Schematic of a panel transducer.

very attractive to vehicles where packaging is a challenge. Finally, depending on the response of the panel on which it is mounted, it can deliver lower frequencies than many small speaker units, which allows more options for sound design and Brand expression. The issue with using PT in this way however is that such a policy needs to be planned and integrated with the infotainment strategy.

9 Conclusion

An approach for designing sounds for EV, both for Interior and Exterior AVAS purposes, has been described. These sounds have been found to radically change the driver and pedestrian perception of the vehicle. The use of NVH Simulation, both interior and exterior, has been found to reduce the development time of these sounds radically by allowing contextual evaluation in the design studio, and enabling collaborative working amongst a geographically distributed team.

Different delivery methods have been evaluated, including Panel Transducers, which were used in this project for AVAS.

References

1. UN Regulation 138, "Uniform provisions concerning the approval of Quiet Road Transport Vehicles with regard to their reduced audibility".
2. ISO 10844:2014, "Acoustics – Specification of test tracks for measuring noise emitted by road vehicles and their tyres".

In-vehicle Analysis of Powertrain-Induced Noise Using FRF Substructures

René Visser[1(✉)], Uwe Fiedler[1], Mladen Chargin[1], and Volker Kreissig[2]

[1] CDH AG, Breitwiesenstraße 19, 70565 Stuttgart, Germany
cdh@cdh-ag.com
[2] Daimler AG, Stuttgart, Germany

Abstract. Numerical analysis plays a significant role in the development of passenger comfort in modern vehicles. Customer perception of quality and comfort is largely influenced by the vibro-acoustic properties of a vehicle. Numerical simulation models of sufficient quality for predicting these characteristics are highly complicated and large numbers of eigenvalues are necessary to enable accurate analysis through the required frequency-range. It follows that extensive computational resources are required and calculation times are long.

This paper describes the generation and use of so-called FRF-Substructures, a modern technique in model reduction for frequency-response analysis in Nastran. Using a practical example of a Mercedes-Benz S-Class model it is shown how FRF-Substructures can significantly increase the efficiency of frequency-response calculations. This enables the evaluation of multiple model variations and complex load-cases in a reasonable time. Further, the advantages of FRF-Substructures are shown by comparison with the traditional superelement approach. Finally, a brief outlook on further uses of the FRF-Substructure technique are addressed.

Keywords: Acoustics · Simulation · FRF-substructuring · Optimization · FEM

1 Introduction

Simulation plays a central role in the development of passenger comfort in modern vehicles. Customers place significant value in the comfort provided by well-engineered acoustic and vibrational behavior of the vehicle. However, the great number of vehicle variants and options present a significant challenge in ensuring that the required level of performance is satisfied in every case. Additionally, the increasing use of hybrid powertrains and electric motors presents further challenges that must be met.

Requirements for vibrational comfort and levels of interior noise are typically defined at full-vehicle level, but are reached by targeted development of vehicle subsystems such as the chassis, powertrain and body. In practice the responsibility for

W. Siebenpfeiffer (Hrsg.): *Automotive Acoustics Conference 2019*, Proceedings, S. 64–70, 2020. https://doi.org/10.1007/978-3-658-27669-0_5

each of these systems is borne by a different team, with each considering the other systems as boundary conditions in their own development work. Simulation models of sufficient quality for this task are highly complicated and high numbers of eigenvalues are necessary to enable accurate analysis through the required frequency-range. FRF-substructuring matches the distributed approach of the development teams and reduces both the required simulation times as well as computational resources.

2 Full Vehicle Example: Analysis of Gear Whining

For the powertrain, the combination of various engine and gearbox options, together with variations such as two or four-wheel drive and the necessary loading scenarios, form a matrix of simulations that cannot be efficiently evaluated using the current standard of vehicle models and model reduction techniques.

Fig. 1. Full vehicle FEM model consisting of both structure and interior air

As an example, shown in Fig. 1, a full-vehicle finite element model for NVH calculations in Nastran can easily consist of over 40 million structural degrees of freedom and a further 1 million for the cavity air model. Goal is to evaluate the interior sound pressure level for multiple powertrain variants subject to gear whining loading. Solving the system of equations in modal space is a well-known technique to save computational time. For such an analysis of gear whine in the Mercedes S-Class, frequency response evaluations are performed up to 1000 Hz. This requires a modal space up to 1500 Hz including an approximate 48000 structural modes and 5000 acoustic cavity modes. On a modern machine, the total calculation time in Nastran is approximately 7.5 h, which can be broken down as shown in Table 1. In this analysis sequence, one cannot identify steps with potential to significantly reduce the computational time.

Table 1. Breakdown of calculation times for the full vehicle analysis

Step	Description	Run time [min]
1	Generating matrices from FE	85
2	Computing structural modes (CDH/AMLS)	225
3	Computing fluid modes (CDH/AMLSF)	5
4	Forming modal matrices	70
5	Solving modal system of equations	55
6	Data recovery	10
Total		450 (7.5 h)

3 Application of Model Reduction Methods

Closer examination of powertrain development reveals that, for a single vehicle variant, simulation models typically feature an identical trimmed body with variations mainly in the powertrain and chassis components. It is therefore possible to reduce the body in the full-vehicle model to representations of the coupling stiffnesses and transfer functions through the body structure and enclosed fluid, while the powertrain and chassis subsystems retain their complexity and flexibility as complete finite-element models.

3.1 External Superelements

Reducing the FE-model of the trimmed body to a standard external superelement [1] in a modal space up to 1500 Hz generates 45000 structural modes. Unfortunately, this is beyond the technical and economical limits of the superelement generation technique even on today's high-performance computers. A large number of modes causes the superelement to have a huge file size and also means the frequency response of the combined job remains costly. Furthermore, superelements do not support acoustic response.

3.2 FRF-Substructures

The "Frequency-Response-Function (FRF) Substructure" reduction technique for Nastran, developed by CDH AG, replaces a portion of a finite-element model with a matrix of transfer functions. These transfer functions represent the dynamic behaviour of a subsystem, here the vehicle body, between the various interface points to the other subsystems and the interior measurement points – including points for acoustic evaluation in the internal cavity (e.g. driver and passenger ears) (Fig. 2).

For this example, the generation of the matrix of transfer functions was completed within 7 h of computation time. This matrix has to be generated only once and can be used for all axle or powertrain variants. The size of the FRF-Substructure matrix

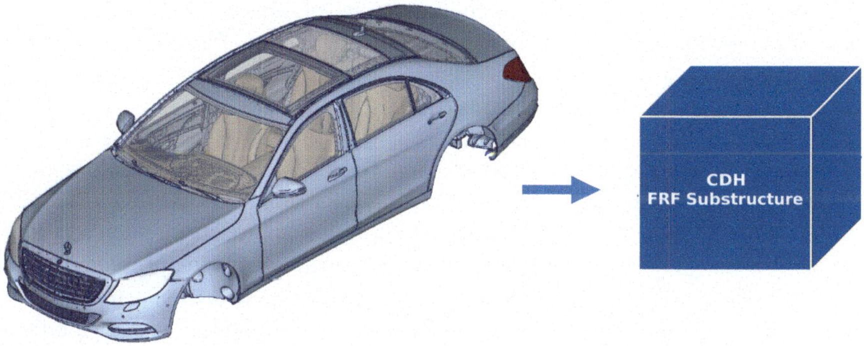

Fig. 2. Reduction of a trimmed body into a FRF-substructure representation

is only determined by the number of frequency steps and the number of coupling, response or excitation DOFs and is not affected by the number of modes.

The FRF-Substructure is coupled with the remaining FE-modelled components (powertrain and chassis) with the result that the calculation time of the full model is reduced from 7.5 h to only 51 min without any loss of accuracy. In fact, compared to the full model, the obtained accuracy can even be further improved since the modal base for the substructure generation can be chosen larger than would be feasible for the analysis of the full FE model.

Fig. 3. Combined model of FRF-substructure and powertrain/chassis (residual structure)

4 Theory of FRF Substructuring

This section presents a brief conceptual overview of the theory behind the FRF-substructuring implementation, for further details reference is made to [6].

For frequency response analysis the equations of motion derived by using the finite element method can be written in the frequency domain form [3]

$$\left[-\omega^2[M] + i\omega[B] + [K]\right][u] = [F] \tag{1}$$

with M the mass matrix, B the viscous damping matrix, K the stiffness matrix and F representing the load vector. Vector u denotes the unknown degrees of freedom (DOFs) that needs to be solved. Generally, the solution approach for such a system of equations consists of solving the generalized undamped eigenvalue problem for a limited number of natural modes. These natural modes, or eigenvectors, can then be used to transform Eq. (1) into the so-called modal space representation with a significantly reduced number of modal DOFs to be solved [2–4]. After solving the reduced modal problem, the data recovery phase transforms modal solution back to the physical DOF vector $[u]$.

As with most model reduction methods, FRF-substructuring involves a two-step approach, the FRF generation step (reduction) followed by the assembly step in which the substructure can be connected to a residual structure. The first step is only done once whereas the second step is usually performed multiple times to evaluate the effect of various design changes in the residual structure.

4.1 Step 1: Generation of FRF-Substructure

The first step in the FRF substructuring involves computation of the transfer matrices of the sub-model that needs to be reduced (here the trimmed body).

To explain the FRF substructuring theory, Eq. (1) is written as:

$$[A][u] = [F] \tag{2}$$

with A being the frequency dependent matrix:

$$[A] = -\omega^2[M] + i\omega[B] + [K] \tag{3}$$

The FRF-substructuring formulation partitions Eq. (2) into 2 parts:

$$\begin{bmatrix} A_{11} & A_{12} \\ A_{21} & A_{22} \end{bmatrix} \begin{bmatrix} u_1 \\ u_2 \end{bmatrix} = \begin{bmatrix} 0 \\ F_2 \end{bmatrix} \tag{4}$$

where part 2 is associated with the substructure its coupling, output and loaded DOFs (referred to as external DOFs) whereas part 1 represents the remaining FRF-substructure internal DOFs. After introduction of the so-called "dynamic flexibility" matrix H that relates the external DOFs u_2 with the corresponding loads F_2

$$[u_2] = [H][F_2] \tag{5}$$

Equation (4) can be written as

$$\begin{bmatrix} A_{11} & A_{12} \\ A_{21} & A_{22} \end{bmatrix} \begin{bmatrix} u_1 \\ H \end{bmatrix} = \begin{bmatrix} 0 \\ I \end{bmatrix} \tag{6}$$

Solving this system of equations for each applied unit load in I yields the complex, frequency dependent and densely populated matrix H that is referred to as the "FRF-substructure". During the substructure generation, residual vectors are applied to the all external degrees of freedom u_2 in order to ensure an accuracy comparable to the original model.

Once H is determined, the unknown external DOFs u_2 of the substructure are computed by a simple and fast matrix vector multiplication for any applied load vector F_2 at every frequency step. However, the inverse operation of determining F_2 from a specified u_2, with H^{-1} referred to as the "dynamic stiffness" matrix, is potentially ill-conditioned, especially when it contains both structure and fluid degrees of freedom. To overcome the ill-conditioned nature of the inverse problem, appropriate stabilizing treatment using regularization techniques is applied [5].

4.2 Step 2: Assembly using FRF Substructure

After generating the FRF-substructure as described in the previous section, it can be connected through its external DOFs to an arbitrary residual FEM model, e.g. the powertrain and chassis as depicted in Fig. 3. The substructure inhibits all the dynamic properties of the original structure but, due to the FRF reduction step, contains only a fraction of the original number of DOFs. Therefore, this technique reduces the overall simulation time and resources drastically without compromising the result accuracy.

Mathematically, any residual FE structure represented by A_{rr} can be attached to the external substructure DOF according to

$$\begin{bmatrix} A_{\mathrm{rr}} & A_{\mathrm{rs}} \\ A_{\mathrm{sr}} & H^{-1} \end{bmatrix} \begin{bmatrix} u_{\mathrm{r}} \\ u_s \end{bmatrix} = \begin{bmatrix} F_r \\ F_s \end{bmatrix} \qquad (7)$$

where A_{rs} and $A_{sr}$are the coupling terms between the residual and the FRF-substructure. As H is frequency dependent, the equation needs to be solved for each frequency step which limits its application in physical coordinates to relatively small systems only. For larger residual models, e.g. like the powertrain and chassis model shown in Fig. 3, the frequency dependent system of Eq. (7) is solved in modal space where the coupling of the physical substructure matrix H to the modal representation of the residual structure is done using a Lagrange multiplier technique. Please note that this section solely describes the simplified conceptual theory of FRF substructuring, the actual implementation contains several more details that are beyond the scope of the current paper.

5 Conclusion

Especially for simulations that involve a high number of modes and models that can be dissected using a limited number of internal connecting points, the FRF substructuring technique gives the engineer an extremely efficient tool to simulate multiple variants in a short computational time. In the current example, the influence of various powertrain options on the acoustic response and other comfort assessments in the vehicle can therefore be investigated more easily and thoroughly.

Apart from the current application, systematic optimization approaches of the structure under demanding load-cases is made viable by the use of the presented substructure technique. A further use case is to enrich a full vehicle model and predict its overall system response with the help of equivalent FRF-substructures that replace parts of which the frequency dependent behavior can only be supplied from an external data source (e.g. measured frequency dependent behavior of a complex rubber mount).

References

1. Bampton C.: Coupling of Substructures for Dynamic Analysis. AIAA Journal. 6 (1968)
2. MSC Nastran, Dynamic Analysis User's Guide, MSC Software (2019)
3. Zienkiewicz O.C.: The Finite Element Method. McGraw Hill Book Company, 3rd ed. (1979)
4. Golub G.H., van Loan C.F.: Matrix Computations. The John Hopkins University Press, 2nd edition (1989)
5. Hansen P.C.: Rank-Deficient and Discrete Ill-Posed Problems. SIAM, 2nd edition (1998)
6. van der Seijs M., de Klerk D., Rixen D.: General framework for transfer path analysis: History, theory and classification of techniques, Mechanical Systems and Signal Processing 68–69, 217–244 (2016)

Measurement of Transfer Matrices of Polyurethane Foam Materials for Sound Absorption and Transmission Loss Applications

Mark Brennan[1(✉)], Rene Boonen[2], Martino Dossi[1], Kristof Verniers[1], and Jan Vandenbroeck[1]

[1] Huntsman Polyurethanes, Everslaan 45, 3078 Everberg, Belgium
mark_brennan@huntsman.com
[2] NABLA Technisch studiebureau, Hauptstrasse 23, 54619 Grosskampenberg, Germany

Abstract. The chemistry of polyurethane flexible foam enables its use in a wide range of sound absorption, sound transmission loss and vibration damping automotive applications. A challenge for developing new acoustic materials is to have a reliable, reproducible and representative small-scale measurement technique to obtain acoustic properties of polyurethane foam. In this paper we present a new measurement technique, based on an impedance tube, where the four pole parameters, or transfer matrices, can be obtained for these materials. A library of transfer matrices for various polyurethane foam materials is obtained and absorption, reflection and transmission coefficients calculated. The materials may also be combined in various layered assemblies. Furthermore, the meaning of the transfer matrix is explored with respect to previous studies on polyurethane foam microstructures.

Keywords: Sound absorption · Sound transmission loss · Impedance tube · Transfer matrix · Polyurethane foam · Foam cell microstructure

1 Introduction

A theoretical model is presented showing how foam cell morphology influences the acoustic properties. The transfer matrix is used to calculate the performance of foam materials in different assemblies. A transmission tube has been developed to measure transfer matrix of samples of absorbing material. The electrical analogy of a sample in a transmission tube can be considered as a T-equivalent circuit of three impedances. Therefore, three measurements with three different closing impedances needs to be carried out to obtain the transfer matrix. From the transfer matrix it is possible to derive acoustic properties of foam materials that agree with the presented theory.

W. Siebenpfeiffer (Hrsg.): *Automotive Acoustics Conference 2019,* Proceedings, S. 71–82, 2020. https://doi.org/10.1007/978-3-658-27669-0_6

2 Theoretical Predictions of the Transfer Matrix for Foam Materials

A plane acoustic wave impinging on a material with a finite thickness can be represented by a transfer matrix such that

$$\begin{bmatrix} p_f \\ v_f \end{bmatrix} = \begin{bmatrix} A & B \\ C & D \end{bmatrix} \begin{bmatrix} p_b \\ v_b \end{bmatrix}$$

where *f* and *b* indicate the forward and backward travelling pressures and velocities of the wave [1]. The transfer matrix depends on material properties and the layer thickness. Knowing the transfer matrix immediately yields the Transmission Loss, *TL*, for the sample:

$$TL = 20 \log \left(\frac{1}{2} \left(A + \frac{B}{Z_0} + CZ_0 + D \right) \right)$$

where Z_0 is a reference impedance, in our case the characteristic impedance of air. It can also be used to calculate reflection and absorption coefficients of different assemblies of materials, the simplest being a sample of material against a hard wall where the overall transfer matrix is

$$[T] = \begin{bmatrix} \mathrm{A} & B \\ \mathrm{C} & \mathrm{D} \end{bmatrix} \begin{bmatrix} 1 & 0 \\ \frac{1}{Z_H} & 1 \end{bmatrix} = \begin{bmatrix} \mathrm{A} + \frac{\mathrm{B}}{Z_H} & B \\ C + \frac{\mathrm{D}}{Z_H} & \mathrm{D} \end{bmatrix}$$

where Z_H is a thousand times Z_0. It follows that the surface impedance of the sample is

$$Z_T = \frac{\mathrm{A} + \frac{\mathrm{B}}{Z_H}}{C + \frac{\mathrm{D}}{Z_H}}$$

and the reflection and absorption coefficients determined by

$$\Gamma = \frac{\frac{Z_T}{Z_0} - 1}{\frac{Z_T}{Z_0} + 1} \qquad \alpha = 1 - \Gamma\Gamma^c.$$

More complex assemblies may also be determined which can include combinations of materials, air gaps, screens etc. [1].

For a fluid layer of thickness d, the transfer matrix is defined as:

$$[T] = \begin{bmatrix} \cos kd & jZ_c \sin kd \\ \frac{j}{Z_c} \sin kd & \cos kd \end{bmatrix}$$

where k is the wave number and Z_c the characteristic impedance of the fluid [1]. Many low-density open cell PU foam layers may be considered as equivalent fluids [2] where the wave number and characteristic impedance are based on a frequency dependent complex effective density and effective bulk modulus according to:

$$Z_c(\omega) = \sqrt{\rho_e(\omega) K_e(\omega)} \qquad k(\omega) = \omega\sqrt{\frac{\rho_e(\omega)}{K_e(\omega)}}.$$

The effective density and effective bulk modulus can in turn be related to the dynamic viscous and thermal permeabilities as follows:

$$\rho_e(\omega) = \frac{j\mu_0}{\omega k_{vis}(\omega)}$$
$$\frac{K_0}{K_e(\omega)} = \gamma + (\gamma - 1)\frac{\rho_0 C_{p,0}}{\kappa_0} j\omega k_{thm}(\omega)$$

where μ_0, ρ_0, $C_{p,0}$, κ_0, K_0 and γ are the viscosity, density, specific heat capacity at constant pressure, thermal conductivity, bulk modulus and heat capacity ratio of air respectively [1]. To calculate the transfer matrix for a foam requires an estimate of the viscous and thermal permeabilities. A method to determine these based on foam morphological characteristics is now proposed.

2.1 Influence of Foam Morphological Properties on the Transfer Matrix

A previous study [3] described a method to construct 3D random periodic polydisperse foam microstructures and used these structures to calculate dynamic viscous and thermal permeabilities of open cell foam microstructures. Various morphologies were studied with varying cell size, polydispersity, porosity and degree of cell opening.

Figure 1 shows pore level simulations, both viscous and thermal, of a particular foam microstructure. The viscous and thermal permeabilities of the foam can be determined from these simulations. To see the effect of morphology on the viscous

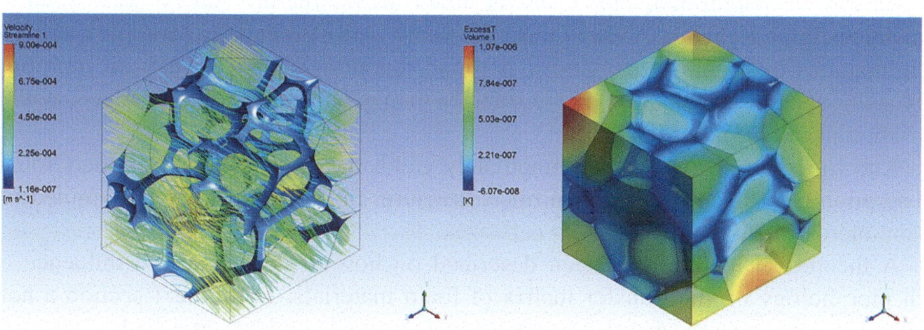

Fig. 1. Pore-level simulations of the air motion (left) and heat conduction (right) of an oscillating wave in an open cell foam from which the frequency dependent viscous and thermal permeabilities can be determined.

and thermal permeabilities a series of 'numerical' experiments must be performed with varying cell sizes, polydispersity, porosity and cell opening.

From the viscous and thermal permeability, it is possible to determine the effective density and effective bulk modulus and ultimately the characteristic impedance and wave number for the foam microstructure according to the equations described in the previous section. An example of how changing the cell size of a particular open cell foam effects the complex characteristic impedance and complex wave number is shown in Fig. 2.

From the characteristic impedance and wave number the elements of the transfer

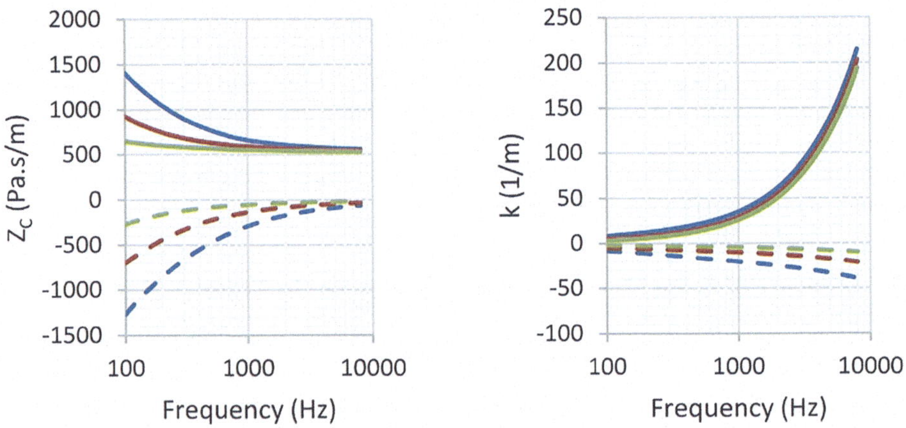

Fig. 2. The influence of cell size (blue = 0.15 mm, red = 0.25 mm, green = 0.5 mm) on the complex characteristic impedance and wave number for a particular foam microstructure.

matrix can be determined, Fig. 3. As the foam microstructure can be considered an equivalent fluid elements A and D are the same. All elements are complex and are combinations of trigonometric and hyperbolic functions. Changes in foam morphology will change each element of the transfer matrix and this is demonstrated, with a variation in cell size, in Fig. 3.

Knowing the transfer matrix immediately yields the Transmission Loss and Sound Absorption Coefficient of the foam microstructure. In Fig. 4 these are determined for foam microstructures with different cell sizes.

A theoretical method has been described on how to determine the influence of cell morphology on the transfer matrix of foam materials. In the next section a new method to determine the transfer matrix of real materials will be described.

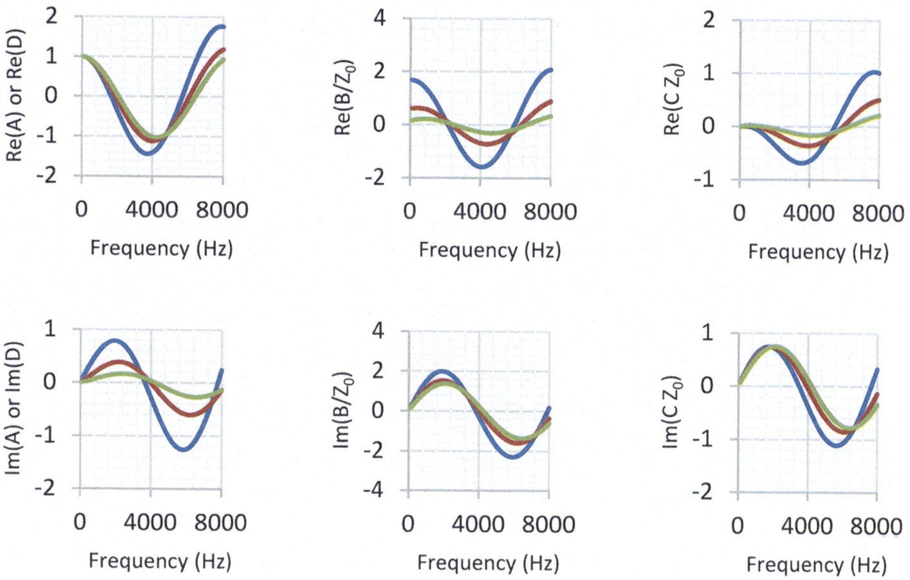

Fig. 3. The influence of cell size (blue = 0.15 mm, red = 0.25 mm, green = 0.5 mm) on the elements (A, B, C and D) of a transfer matrix of a particular foam microstructure with a layer thickness of 3 cm.

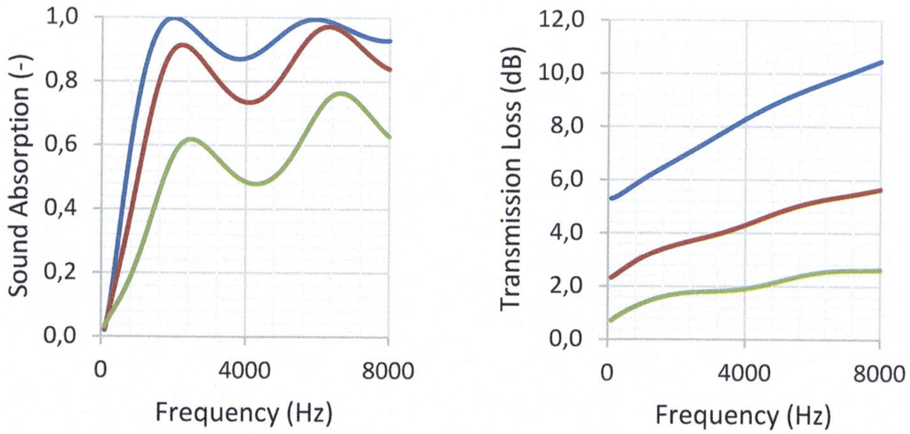

Fig. 4. The influence of cell size (blue = 0.15 mm, red = 0.25 mm, green = 0.5 mm) on the sound absorption coefficient and transmission loss, determined from the transfer matrix, of a particular foam microstructure with a layer thickness of 3 cm.

3 Measurement Set-up of Transmission Tube

A new method to measure the four-pole matrix of an absorbing material in a transmission has been developed [4]. A scheme of the transmission tube is shown in Fig. 5. In this tube, a second microphone arrangement has been introduced between the sample impedance Z_S and the closing impedance Z_A. The measurement head with microphones 1 and 2 measures the impedance behind the first reference section, consisting of the sample impedance, the impedance of the second measurement head and the closing impedance. The measurement head with microphones 3 and 4 measures the impedance behind the second reference section, consisting of only the closing impedance Z_A.

Using an electrical analogy, and from electrical circuit theory, it is known that any passive linear two port circuit can be represented by a T-equivalent circuit according to Thévenin's theorem [5]. The sample can therefore be represented by three impedances Z_1, Z_2 and Z_3, Fig. 6.

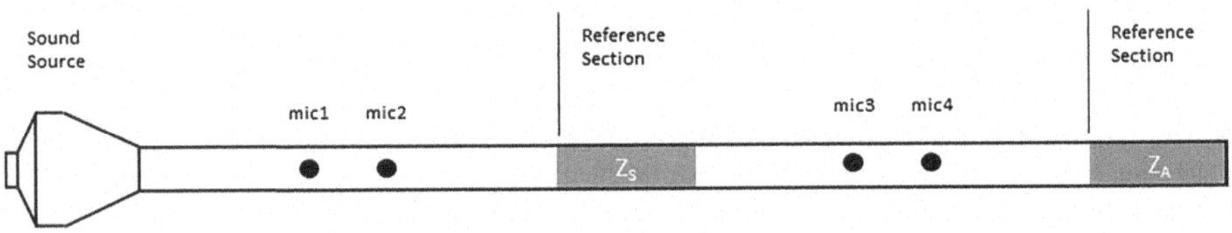

Fig. 5. A transmission tube with sample Z_S, closing impedance Z_A and two microphone measuring heads.

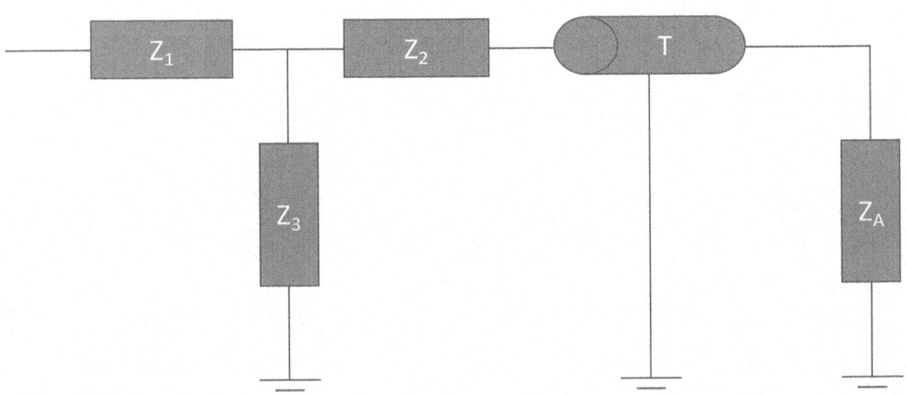

Fig. 6. Electrical equivalent circuit of the sample Z_S (represented by Z_1, Z_2 and Z_3), the part of the transmission tube with the second measurement head, *T*, and the closing impedance Z_A.

To determine the three impedances Z_1, Z_2 and Z_3 three impedance measurements at each measurement head must be carried out with three different closing impedances using the ISO 10534-2 two microphone transfer function method. Ideal choices for the closing impedances would be a hard wall termination, $Z_A = \infty$, open termination, $Z_A = 0$ and an anechoic termination, $Z_A = Z_0$. However, these ideal impedances are not necessary and as long as the three termination impedances are sufficiently different it will be possible to determine Z_1, Z_2 and Z_3.

From these measured impedances, a set of three algebraic equations will be set up from which Z_1, Z_2 and Z_3 will be solved in amplitude and phase. The transfer matrix can be determined from Z_1, Z_2, Z_3 and Z_0, the characteristic impedance, as follows (Fig. 7):

$$[T] = \begin{bmatrix} 1 + \frac{Z_1}{Z_3} & Z_0\left(Z_1 + Z_2 + \frac{Z_1 Z_2}{Z_3}\right) \\ \frac{1}{Z_3 Z_0} & 1 + \frac{Z_2}{Z_3} \end{bmatrix}$$

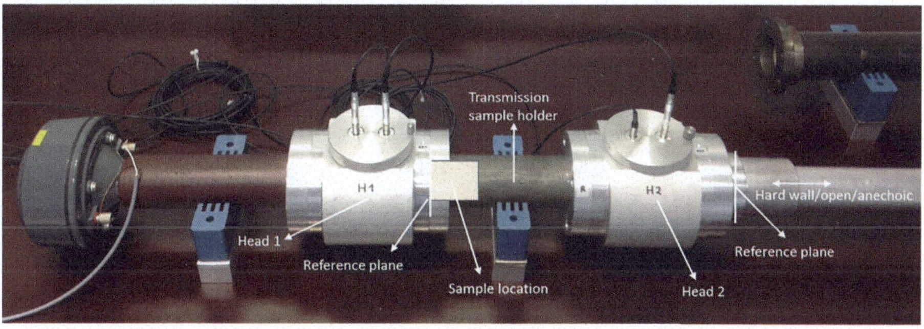

Fig. 7. Transmission tube with internal diameter of 4.5 cm with two measurement heads and a closing impedance.

4 Experimental Results and Comparison with Theory

Figure 8 shows three terminations impedances, measured according to ISO 10534-2 two microphone transfer function method representing hard-like termination, open termination and anechoic-like termination. With these three impedances it is possible to obtain the T-equivalent circuit impedances, Z_1, Z_2 and Z_3, for a foam sample, Fig. 9. The two series impedances Z_1 and Z_2 are similar and behave like ideal acoustic capacitors with a phase angle of $-90°$ and a magnitude drop of 20 dB/decade. The impedance Z_3 behaves more like an non-ideal acoustic inductor with a phase angle of $90°$ and a frequency dependent resistance element.

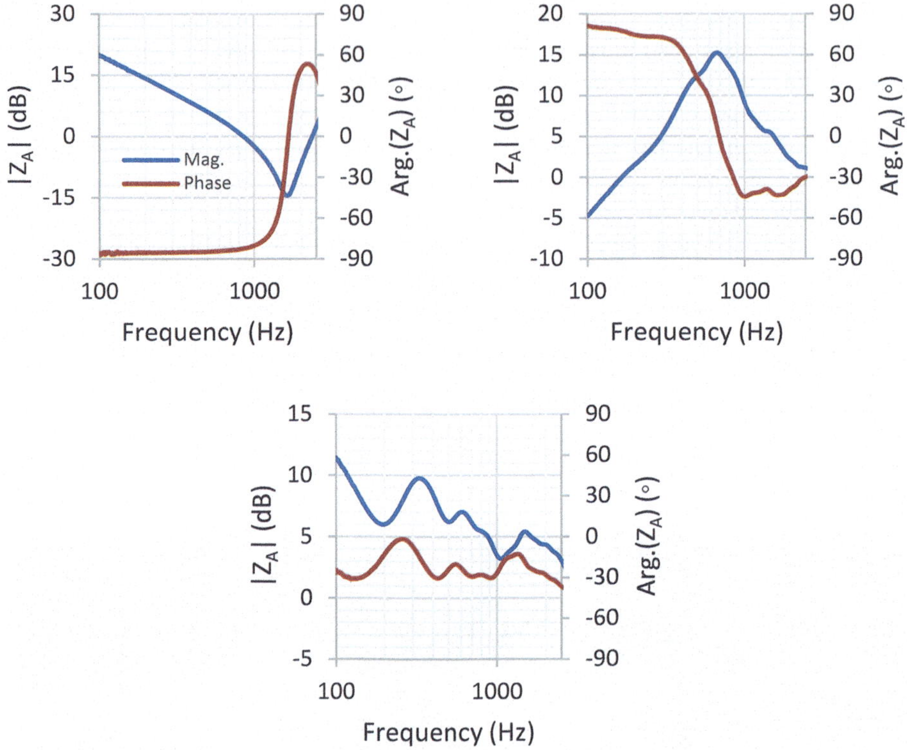

Fig. 8. Examples of three termination impedances, Z_A, sufficiently different to be able to calculate the impedances Z_1, Z_2 and Z_3 where the reference 0 dB is Z_0.

Knowing impedances Z_1, Z_2 and Z_3 it is possible to calculate the elements of the transfer matrix. In Fig. 10 are the elements of the transfer matrix for two PU foams of different densities. Both sets of elements behave in a similar manner to what would be expected from an equivalent fluid type material, that is that only the foam cell morphology is important for predicting acoustic properties of these two foams. Notice that element A and D are the same for both foams, confirming the equivalent fluid hypothesis.

As the fluid equivalent assumption holds, it is possible to derive the complex characteristic impedance and wave number of the material from the elements using the following relations:

$$Z_c = \sqrt{B/C} \quad k = \frac{1}{d}\tan^{-1}\left(\frac{Re(\sqrt{-BC})}{Re(A)}\right) + \frac{j}{d}\tanh^{-1}\left(\frac{Im(A)}{Re(A\sqrt{-BC})}\right)$$

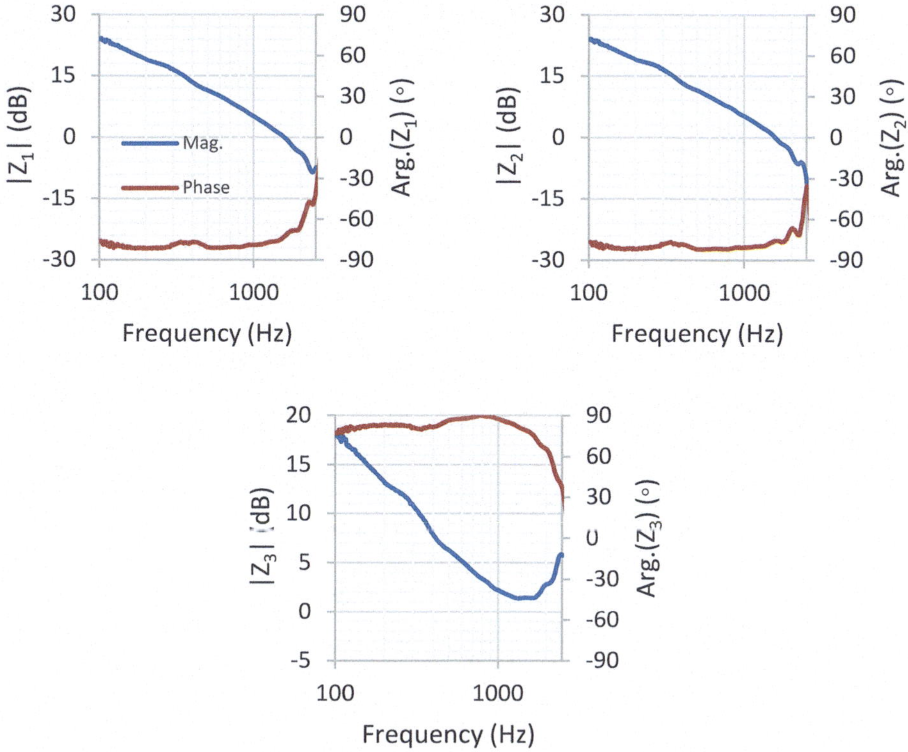

Fig. 9. Measured impedances Z_1, Z_2 and Z_3 from the T-equivalent circuit where the reference 0 dB is Z_0

where d is the sample thickness. These properties are independent of sample size and are shown in Fig. 11 for the two PU foams of different density. Knowing these parameters, it is now possible to calculate the acoustic properties in different assemblies. Two examples are shown in Fig. 12, the Sound Absorption Coefficient calculated for the foams against a rigid backing and the Sound Transmission Loss through the foam materials.

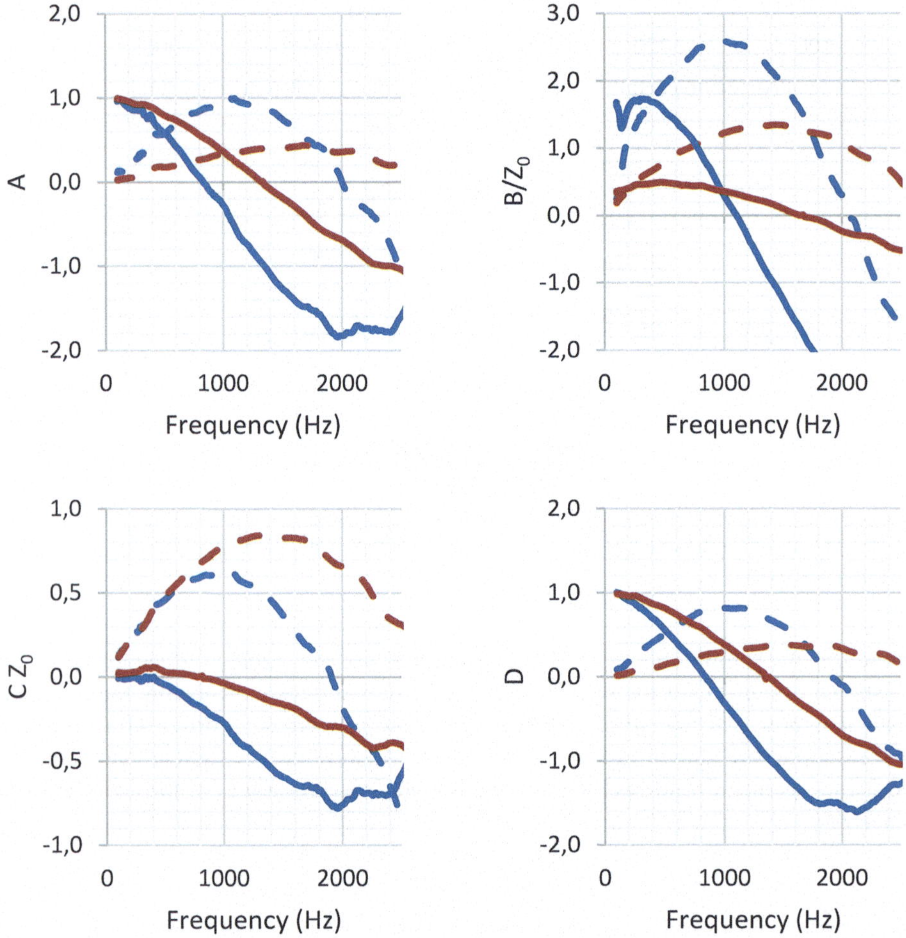

Fig. 10. The elements (A, B, C and D) of the transfer matrix of two PU open cell foam samples with densities 100 kg/m^3 (blue) and 40 kg/m^3 (red) and a thickness of 5 cm.

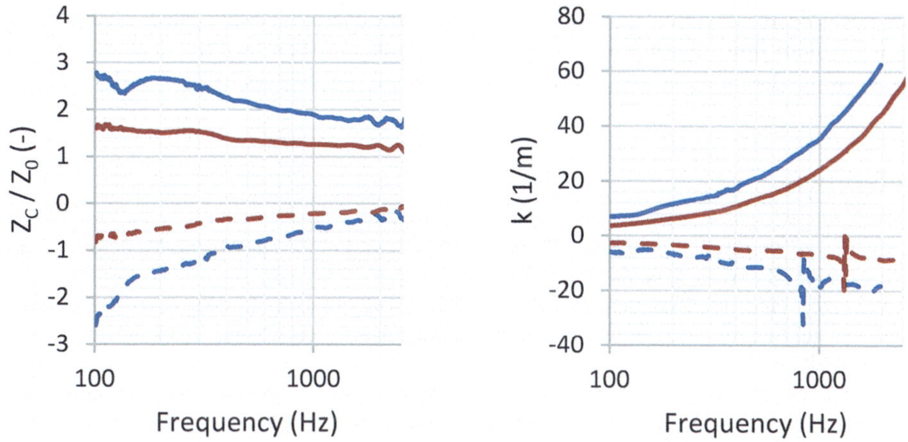

Fig. 11. The complex characteristic impedance and wave number, derived from the transfer matrix, of two PU open cell foam samples with densities 100 kg/m^3 (blue) and 40 kg/m^3 (red).

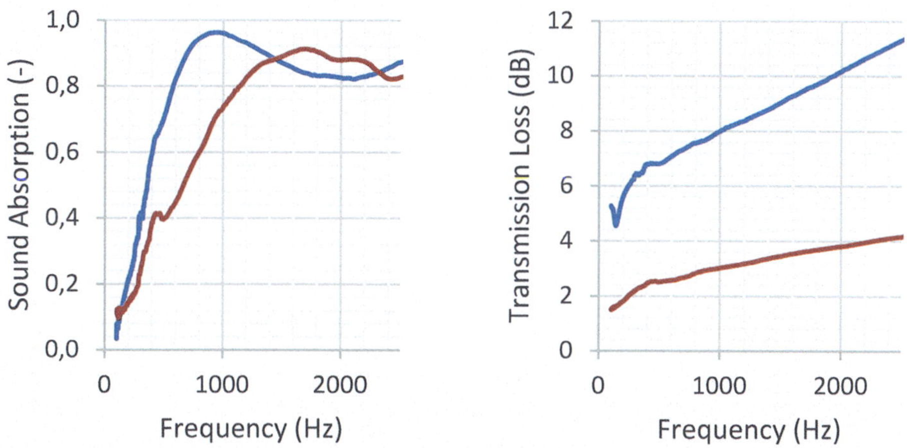

Fig. 12. The sound absorption coefficient and transmission loss, derived from the transfer matrix, of two PU open cell foam samples with densities 100 kg/m^3 (blue) and 40 kg/m^3 (red) and a sample thickness of 5 cm.

5 Conclusion

A theoretical method for calculating the transfer matrix of an open cell foam was presented and it was shown how changes in the foam cell morphology can influence the transfer matrix. A new method for determination of the transfer matrix by impedance measurements was also presented. The sample is represented by a T-equivalent circuit of three impedances. These three impedances can be identified from three impedance measurements with three different closing impedances. From these measurements it was possible to calculate the transfer matrix. The example foams measured confirm the equivalent fluid hypothesis for open cell foams and therefore the acoustic properties, namely, the characteristic impedance and wave number could also be derived.

References

1. Allard, J.-F. and Atalla, N., (2009) Propagation of Sound in Porous Media: Modelling Sound Absorption Materials, London: Wiley
2. Dossi, M., Huo, L., Brennan, M., Moesen, M., Vandenbroeck, J.; "An Inverse Method to Determine Acoustic Parameters of Polyurethane Foams", InterNoise 2019, Madrid, Spain (June 2019)
3. Brennan, M., Joncheray, T., Vandenbroeck, J.; "A numerical method to study the influence of the three dimensional geometrical features of porous polymeric foam on its sound absorbing behavior", 1st Automotive Acoustics Conference, Zurich, Switzerland (July 2011)
4. Boonen, R.; "Measurement of the Four Pole Matrix of a Sample in a Transmission Tube", DAGA, Rostock, March 2019
5. Beranek, L. and Mellow, T.; "Acoustics: Sound Fields and Transducers", Oxford: Elsevier

Attribute Balancing – Trading-Off Noise Versus Emissions, Ride and Handling and Driveability

Simon Tate[1], Ben Munday[2], Phil Crowther[2], and Matt Maunder[1](✉)

[1] Ricardo UK Limited, Shoreham-by-Sea BN43 5FG, England
matthew.maunder@ricardo.com
[2] Ricardo UK Limited, Leamington Spa CV31 1FQ, England

Abstract. To develop optimum vehicle designs, engineers must balance vehicle attributes to meet various product requirements, relating to specific market demands. It is commonplace to see trade-off relationships between noise and other important attributes. Balancing decisions are best made when detailed information is available through appropriate simulation and testing. This paper describes three case studies exemplifying such trade-off situations, and the technical approaches taken to decide the best attribute balance for each vehicle product.

Demand is increasing for outstanding fuel economy in gasoline engines especially in Hybrid Electric Vehicles (HEVs). An enabler for good efficiency is rapid combustion, but this can erode the refined sound quality that customers expect from gasoline engines. A simulation approach was used to improve gasoline combustion noise early in a project. Mitigation measures were optimized for drive-cycle fuel efficiency, combustion noise and engine mass.

Car makers must provide cruising comfort through low road noise (more prominent with quieter electric propulsion) and simultaneously deliver competitive driving dynamics. A vehicle structure's dynamic response strongly influences both ride comfort and road noise, and is significantly impacted by the integration of a heavy battery pack. Using a multi attribute optimization approach, an optimum balance was found for a Battery Electric Vehicle (BEV) to control road noise whilst giving good vehicle structural response and reduced mass.

Electric cars deliver strong and responsive vehicle performance and good cruising comfort, but their dynamic driving experience and brand appeal is marred by the lack of informative acoustic feedback from the power unit. An augmented power unit sound system was calibrated to deliver an exciting and authentic acceleration feel during spirited driving, while retaining acoustic comfort during cruising and minimizing environmental noise pollution.

Keywords: Multi-attribute optimization · Noise · Computational fluid dynamics · Spark ignition · Combustion noise · Structural dynamics · Road noise · Augmented sound

W. Siebenpfeiffer (Hrsg.): *Automotive Acoustics Conference 2019,* Proceedings, S. 83–98, 2020. https://doi.org/10.1007/978-3-658-27669-0_7

1 Introduction

In automotive organizations, the development of the Noise, Vibration and Harshness (NVH) is typically the responsibility of a team of experts. This team may sit within one department spanning the complete vehicle and its many systems, or sometimes split within several separate groups specializing in systems such as engine, chassis, transmission and driveline etc. To achieve excellence in vehicle NVH for the customers, the NVH team's responsibility is to balance the noise and vibration contributions from these different systems to achieve good overall performance in interior or exterior NVH performance. Analysis tools and methods for balancing these different contributions have been used extensively for many years [1].

To develop optimum vehicle designs, the automotive organisations must balance all vehicle attributes to meet various product requirements. In many instances, NVH performance will trade off against other vehicle attributes such as efficiency, dynamics and performance feel. Engineers are therefore required to co-operate between different functional teams to tune the optimum attribute balance relating to specific market demands.

Balancing decisions are best made when detailed information is available. A request to trade off attributes is unlikely to be accepted without a thorough justification which shows the expected NVH benefit and the magnitude of any tradeoff. It is also beneficial for this balancing activity to occur early in the product development cycle, before the designs of software and hardware are sufficiently progressed to make any changes difficult and expensive to implement [2]. The ability to quantify in a timely way this attribute balance can be provided through appropriate simulation and testing. This paper describes three case studies exemplifying such trade-off situations, and the technical approaches taken to decide the best attribute balance for each vehicle product.

2 Gasoline Combustion Noise Development

2.1 Background

The light duty certified engine market has been experiencing major changes in the technology mix over recent years. One major driver is global air quality concerns, particularly nitrous oxide (NOx) and particulate matter (PM) emissions from diesel engines. Concern over the suitability of diesel engines, particularly in cities, is driving a reduction in the sale of diesel-powered vehicles [3].

By 2020, European fleet average carbon dioxide (CO_2) targets are set to be reduced by 27% compared to the 2015 average levels [4] with similar CO_2 reduction targets being introduced globally. Stagnation in the reduction of fleet average CO_2 can be attributed to the increase in gasoline market share plus a rise in the popularity of SUV vehicles which are typically larger (with higher drag and rolling resistance) than the vehicles they are replacing. Engine downsizing has previously been successful in reducing engine friction from reduced cylinder count, but real-world fuel economy has not necessarily improved due to the typical operating conditions in the field. These factors present an increasing challenge for the vehicle manufacturers in meeting holistic CO_2 targets.

In response, gasoline strategy is being revisited. The focus has been shifted to one of very high Brake Thermal Efficiency (BTE). This strategy is being coupled with powertrain electrification in Hybrid Electric Vehicles (HEV's) and automatic transmission calibrations, where the engine can maintain larger periods of residency at these points of high thermal efficiency.

Studies of gasoline combustion [5] have shown that high tumble rates are required to meet BTE targets of >40% and a consequence of the high tumble rates is fast combustion. Ricardo benchmark analysis has shown a historical trend for increase in peak gasoline combustion rates (see Fig. 1) which has been driven by increase in specific performance but is also more recently due to the increase in-cylinder air motion and elimination of high-load fuel enrichment. The increased peak pressure rise rates are leading to the generation of significant levels of gasoline combustion noise (the maximum rate of pressure rise in-cylinder is a common and appropriate gasoline combustion noise metric). In addition to the peak levels at high load, higher pressure rise rates are also being seen at part load key points targeting maximum levels of thermal efficiency. At these conditions, vehicle speeds and operating loads are often low, which reduces the levels of background mechanical or road noise which may otherwise mask the combustion noise. Figure 2 shows the spectral content of the cylinder pressure a modern turbocharged direct injection gasoline engine compared to a modern diesel from the Ricardo database at an identical part load condition. It is noted that at frequencies above 1 kHz,

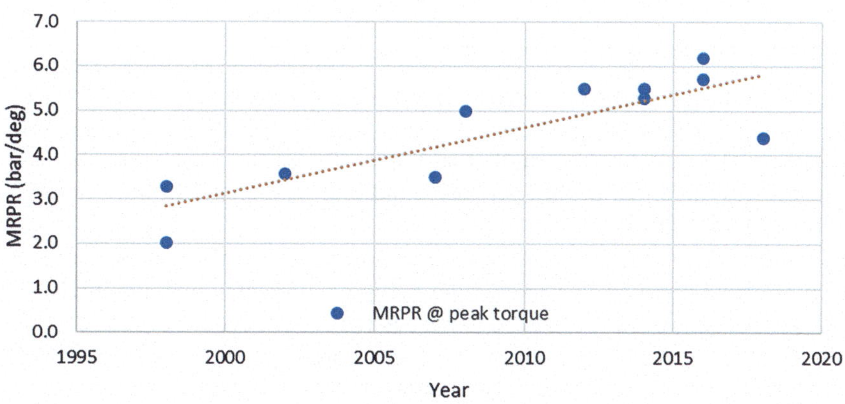

Fig. 1. Progression of gasoline engines' Maximum Rate of Pressure Rise (MRPR) at peak torque condition against year of test from Ricardo SONAR benchmark database

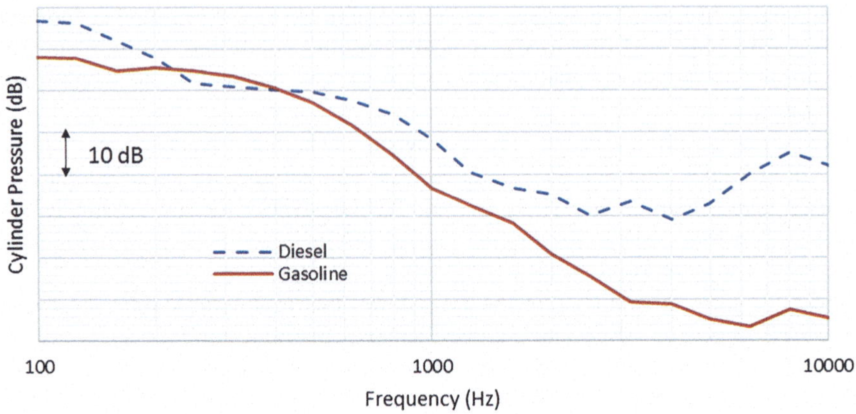

Fig. 2. Comparison of cylinder pressure spectra of modern diesel and gasoline engine for an identical part load condition

diesel cylinder pressure levels are much higher than gasoline. However, at frequencies below 1000 Hz, the cylinder pressure is more comparable. Ricardo has typically seen combustion noise issues in gasoline engines in the 400–1000 Hz region at part load conditions.

A case study is presented where Ricardo used a simulation and test approach to rapidly investigate a concern with gasoline combustion noise. Mitigation measures were optimized for acceptable attribute tradeoffs on drive-cycle fuel efficiency, combustion noise and engine mass.

2.2 Gasoline Combustion Development Case Study

A vehicle manufacturer requested that Ricardo investigate a subjective NVH concern on one of their early prototype vehicles which was described as a "rough" vehicle sound quality. The vehicle was installed with a small gasoline engine and the subjective concern was identified in the mid-speed and mid-load area following initial calibration for RDE/WLTP cycles.

The engine was installed by Ricardo in a powertrain semi-anechoic test facility and the powertrain noise levels and cylinder pressures measured over the full load and speed range. It was noted that the load cases where the sound quality issue was identified corresponded to relatively high levels of maximum rate of pressure rise (MRPR) and high noise levels in the 400–800 Hz frequency band.

The speed-load region of the calibration map where the NVH concern was found is shown highlighted in Fig. 3. It should be noted that this area of the map did not see the highest MRPR, but the combustion noise was high relative to masking from other sources. Techniques such as combustion mechanical breakdown [6] can be used to set targets for combustion noise contribution as a percentage of total noise, but in this project, it was decided to use MRPR as a metric for combustion noise as this is commonly measured parameter by the calibration team and Ricardo experience has shown it to correlate closely with combustion noise for gasoline engines. Targets for MRPR were therefore cascaded for nine key points in the region where the concern was identified, using data from best in-class powertrains from the Ricardo SONAR NVH database. The targets included consideration of peak level in this area of the map and provided suitable progression of increased combustion noise with increased speed and load.

	Speed /Rev/min										
	1000	1500	2000	2500	3000	3500	4000	4500	5000	5500	6000
Min Load ↑	1	1	1	1	1	1	1	1	1	1	1
	2	2	2	2	2	2	2	2	2	2	2
	4	3	3	3	3	3	3	3	3	3	3
	5	4	4	4	4	4	4	4	4	4	4
	5	5	5	5	5	5	5	5	5	5	5
	6	5	7	6	6	6	6	6	6	6	5
	6	5	7	7	7	7	6	5	5	6	5
	6	6	7	6	7	6	5	5	5	5	4
	6	6	5	6	7	6	5	5	5	4	4
	6	6	5	6	6	5	5	4	4	4	3
	5	5	5	5	5	5	5	4	4	3	3
Max Load ↓	5	5	5	5	5	5	5	4	4	3	3

Fig. 3. Measured calibration key point map of MRPR (bar/°) for gasoline calibration with highlighted area showing combustion noise concern

To reduce testing costs and shorten development times, Ricardo utilized an existing one-dimensional Computation Fluid Dynamic (CFD) model in Ricardo WAVE software to predict MRPR. One-dimensional CFD simulations are commonly used in performance and emissions predictions and can provide MRPR predictions with minimal additional effort. Previous studies [7] have shown that three-dimensional CFD models are required for diesel combustion noise predictions. This is because combustion chamber resonances have a strong influence in the frequency range typically seen for diesel combustion noise issues and 3D modelling will therefore capture this geometry. As discussed previously, the frequency range with high levels of excitation is below 1 kHz in gasoline engines. Combustion chamber resonances are typically above 5 kHz in gasoline engines, so a one-dimensional model can be used with acceptable accuracy for the cylinder pressure calculations without the extensive modelling effort and simulation durations required to capture the three-dimensional behavior. A reasonable correlation was achieved with the WAVE model and test data for the baseline calibration as shown in Figs. 3 and 4.

	Speed /Rev/min										
	1000	1500	2000	2500	3000	3500	4000	4500	5000	5500	6000
Min Load ↑	1	1	1	1	1	1	1	1	1	1	1
	2	2	2	2	2	2	2	2	2	2	2
	3	3	3	3	3	3	3	3	3	3	3
	5	5	5	5	5	4	4	4	4	4	4
	6	6	6	6	6	5	5	5	5	5	5
	7	7	8	8	7	6	6	6	6	6	5
	6	7	8	8	8	7	6	6	6	6	5
	5	6	7	8	8	7	5	5	5	5	5
	5	5	6	7	7	6	5	4	4	4	5
	5	5	6	6	7	6	5	4	4	4	4
	4	5	5	6	6	5	5	4	4	4	4
Max Load ↓	4	4	5	5	5	5	5	4	4	4	4

Fig. 4. WAVE-modelled calibration key point map of MRPR (bar/°) for gasoline baseline calibration with highlighted area showing combustion noise concern

Parametric iterations of the engine were performed using the WAVE model to identify potential changes to the engine calibration to improve the MRPR in the concern region. Example parameters considered included spark retard, camshaft timing and injection timing changes. In this study, manual iterations were performed to reduce the MRPR at these load cases using the Ricardo benchmark database as a guideline for acceptable levels. For improved NVH, a reduction of 1–2 bar/° at each key point was achieved using the parametric iterations in the model.

In parallel with the MRPR analysis, a vehicle systems simulation using Ricardo IGNITE software was performed. The analysis utilizes the engine fuel consumption maps from the WAVE model, the drive cycles (e.g. WLTP) and the vehicle system data such as gear ratios and vehicle inertia. An overview of the scope of the vehicle system model is shown in Fig. 5.

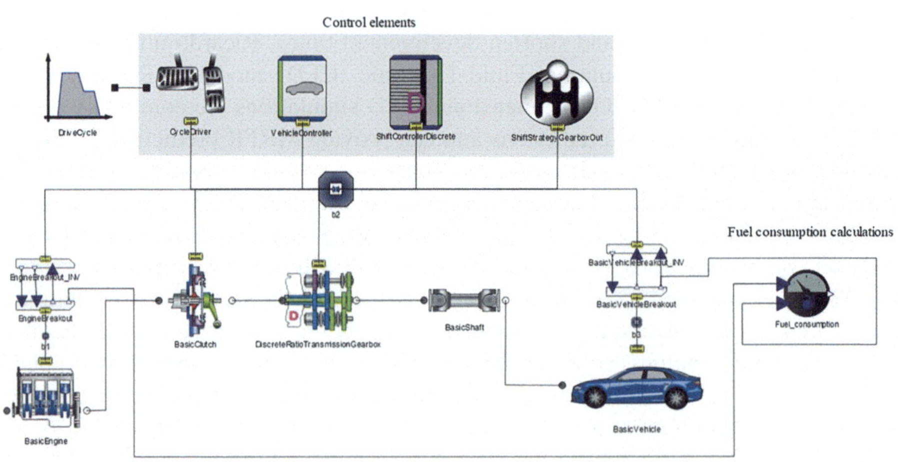

Fig. 5. Overview of vehicle systems modelling

The model predicted a trade-off penalty of 0.3% fuel consumption over the prescribed legislative drive cycle from the NVH optimization. In addition to the NVH concern area of the speed and load map, the transition of the calibration strategy was investigated to ensure the changes were ramped in gently over a wider region without sudden step change. The size of this transition region was iterated using the models and optimized to a point were no further trade-off to fuel consumption was required. The vehicle was predicted to meet target fuel consumption with the changes and these were considered preferable to additional cost and weight penalties from acoustic palliatives such as vehicle or powertrain noise shields.

These changes were then validated on the NVH testbed and drive cycle emissions, to ensure that overall requirements were met. Figures 6 and 7 show the MRPR for baseline and revised calibration at nine test key points which provide the required NVH improvements. MRPR improvements were typically in the range 1–2 bar/° and gave in excess of one point subjective improvement on a ten point scale.

	Speed (Rev/min)		
Load	1750	2000	2250
30%	5	5	5
40%	6	6	6
50%	6	7	6

Fig. 6. Measured NVH key point MRPR (bar/°) for gasoline baseline

	Speed (Rev/min)		
Load	1750	2000	2250
30%	4	4	5
40%	4	4	5
50%	5	6	6

Fig. 7. Measured NVH key point MRPR (bar/°) for gasoline NVH optimization

3 Road Noise and Vehicle Structure Development

3.1 Cross Functional Battery Pack Integration

There are a range of critical issues which impact the market attractiveness of BEVs ranging from unit price, vehicle range and driving attributes to infrastructure availability. This work focused on strategies for optimizing the vehicle mass whilst giving good dynamic performance – particularly with respect to road noise. A key concern for BEVs is interior road noise, which is increasingly significant due to reduced masking from other sources and increasing excitation from low rolling resistance tires. Structural performance of the battery pack both before and after it is fitted to the vehicle is key to ensure high voltage safety from crush and crash events whilst maximizing energy storage and therefore vehicle range. Since the battery pack requires a high degree of structural integrity as a standalone component, it can also significantly contribute to the vehicle's overall structural integrity. Optimizing battery structure and mounting for the vehicle application is therefore important to achieve a low total vehicle mass.

A case study is presented where Ricardo performed an in-vehicle structural optimization of a battery pack, considering multiple attributes (crush resistance, static and dynamic stiffness and road noise), for a luxury sedan vehicle. This provided strategies for optimizing the vehicle mass whilst giving good dynamic performance, particularly with respect to road noise.

3.2 Baseline BEV Vehicle

The donor vehicle for the study was an E segment combustion engine sedan. An initial design for the battery pack's frame, cover panels and modules were created using a parametric model allowing the setting of profiles for different battery frame sections. Starting parameters were selected, based on experience, to meet crush requirements and to give a good trade-off between crush and stiffness in torsion and bending. L-shaped sections were chosen for the battery frame for packaging efficiency and connection to the Body in White (BIW).

3.3 Model Preparation

The applicable crush test involves a 150 mm diameter rigid pole moving in the horizontal plane (longitudinal and lateral directions) [9] – a crush force of 200 kN is required without intrusion-generated fire or explosion risk. The crush loading is highly localized, allowing the use of smaller models of local areas of the pack and so increasing the efficiency of the simulation.

To evaluate the stiffness of the baseline BEV vehicle, the model was excited vertically in phase at both sides of the vehicle at the B-pillars to induce vertical bending, and excited vertically out of phase at the front damper top attachment points to induce torsion. Natural modes were calculated but not used in the optimization process – they were assessed manually.

For the road noise analysis, the trimmed body CAD model was meshed along with the acoustic cavity within the cabin. To reduced calculation time, a structure-acoustic super-element of the trimmed body and its acoustic cavity was used. The super-element retained sufficient degrees-of-freedom to allow for connection of the battery pack and application of excitation to the vehicle super-element. The model was excited through the wheel hubs using principal component forces extracted from measured road noise data on the donor vehicle using a matrix inversion method. Sound pressure results for road noise were then extracted from the model at the response points.

3.4 Sensitivity Analysis

To identify the most suitable parameters for the optimization, a sensitivity study was performed on the model considering the effect on the vertical, horizontal and torsional modes of the body of material thickness and number of connections to the battery. It was found that for the baseline model, the number of connection points had very little influence on the body modes. With the introduction of three additional floor cross members to the BIW at the battery pack connection points, to increase local stiffness, sensitivity to the number of connection points increased.

A further study was performed, evaluating the effect of battery module stiffness on the free-free modal results of the battery pack. The starting assumption for this study was non-structural battery modules, with the battery modules' first mode set to be in the range 35–45 Hz. The frequency range for the battery modules was chosen to position their response away from vehicle fundamental modes and secondary ride frequencies, within the constraints of practical module mounts. This initial study was carried out to understand what value of the Young's Modulus should be used for the simplified representation of the battery modules fitted in the battery pack model. From the results (see Fig. 8) it was decided to use a Young's Modulus of 100 MPa for the

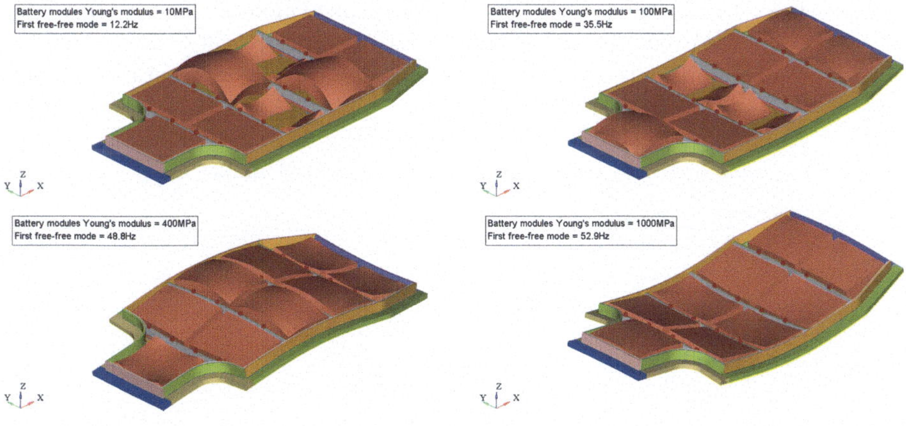

Fig. 8. Modal responses of battery pack structure with different Young's Moduli

optimization, thus predicting a first mode at 35.5 Hz. It was observed that stiffness of the battery modules influenced free-free modes of the battery pack significantly.

3.5 Multi-attribute Optimization

The objective of the optimization was to minimize mass, subject to constraints on stiffness, first mode, noise and crush as shown in Table 1.

Table 1. Optimization constraints

Optimization constraints	Min	Max
Front crush reaction force	200 000 N	
Rear crush reaction force	200 000 N	
Side crush reaction force	200 000 N	
Bending stiffness	11.1 kN/mm	
Torsion stiffness	25.2 kN/°	
Overall frequency response (driver's ear)		46.12 dBA

The following optimization variables were selected: the material of the frame, the number of connection points between the body in white and the battery pack, and the geometry of the battery frame. 15 design variables were identified for the frame geometry, comprising height, width and thickness of the bottom and height and thickness of the top for each of the front, rear and side sections. For the attachment points, a 0.5 kg mass penalty was assigned for each additional attachment.

The optimization process is illustrated in Fig. 9. Approximation models were generated in advance of optimization through random sampling followed by application of response surface methods. The approximation models allowed rapid assessment of crush performance during the optimization. The response surface models predicted crush performance within 5% of the performance predicted by the FE model, with outliers up to 10%, allowing fast identification of unfeasible designs. For designs with crush resistance below target, models with longer calculation times (modal, stiffness and NVH) were not executed – their outputs were assigned penalty values. This time saving allowed significantly more design parameters to be assessed in the optimization process.

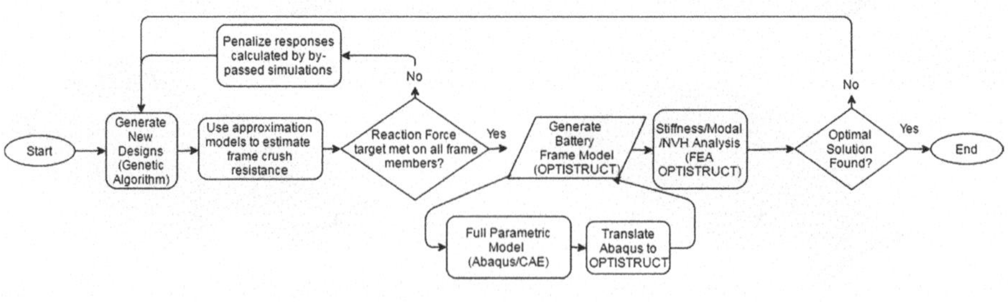

Fig. 9. Optimization process flow chart

The large number of variables and the non-linearity of the responses in the study required the use of a global search optimization, due to the difficulty in selecting an appropriate starting point for a local algorithm. A Multi-Island Genetic Algorithm (MI-GA) was selected as the optimizer. This algorithm performs the traditional genetic operations, but on a population that is spread across multiple islands. Migration of the population between the islands occurs every few generations. It has the advantage of being executable in parallel. It was set up to consider 10 islands, with 10 individuals each, a mutation rate of 0.025 and migration occurring every third generation. Once the genetic algorithm was judged to have found the global optimum zone (Fig. 10), a local search algorithm was used to continue the search for the global optimum. A Downhill Simplex method was used for this purpose, with the best design found by the MI-GA providing the starting point for the search.

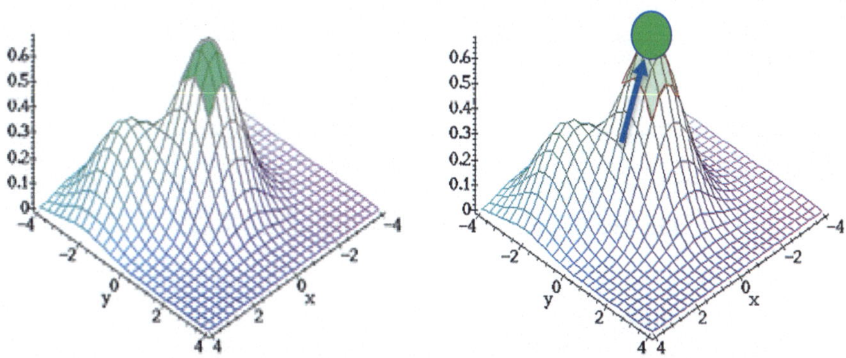

Fig. 10. Global optimum zone (left) and global optimum point (right)

3.6 Outcomes

Adding the battery pack to a gasoline vehicle body-in-white increased the vehicle mass by 633 kg (see Fig. 11) – the 582 kg battery being very similar in weight to that of the Tesla Model X P90D battery is 580 kg [8]. Optimizing the battery pack stiffness and vehicle integration offered a potential 80 kg reduction whilst still meeting crush requirements and giving competitive road noise performance. Attaching the stiff battery package increased the static stiffness of the vehicle by 35% (torsion) and 41% (vertical bending). The first torsional mode frequency was unchanged, and the first vertical bending mode frequency reduced by 27%. The overall sound pressure was unchanged and the response around 40 Hz was improved by the optimization, linked to revised modal matching and improved floor stability. The optimization process reduced the battery pack frame mass by 68% from 117.3 kg to 37.5 kg.

The optimized design incorporates an aluminum battery frame with 14 fixings. The stiffness of the frame is critical to the design performance as is the stiffness of the vehicle structure at the battery attachment points. Further investigation showed that the number of battery fixings had a negligible effect above a total of 14, but optimum fixing positions on the body side were important. Partial optimization suggests that

		Results	Units	Baseline	Initial (Steel)	% Increase	Optimised (Aluminium)	% Increase	
Body in White (BIW)	Mass	BIW mass	kg	455.5	499.6	9.7%	499.6	9.7%	0%
		Battery pack mass	kg	0.0	582.3		501.9		-14%
		Battery pack to BIW connection mass	kg	0.0	7.0		7.0		0%
		Sum Mass	kg	455.5	1088.9	139.1%	1008.5	121.4%	-7%
	Modal	Torsion mode	Hz	41.8	41.4	-0.9%	41.1	-1.7%	-1%
		Vertical bending mode	Hz	49.3	39.2	-20.5%	35.8	-27.4%	-9%
		Horizontal bending mode	Hz	48.5	47.2	-2.6%	46.0	-5.1%	-3%
	Stiffness	Body torsional stiffness	kNm/deg	25.2	35.2	39.3%	34.1	35.3%	-3%
		Vertical bending stiffness	kN/mm	11.1	16.2	45.7%	15.8	41.9%	-3%
Trimmed BIW		Overall Sound Pressure Level	dB(A)	46.12	46.64	1.1%	46.13	0.0%	-1%
	NVH			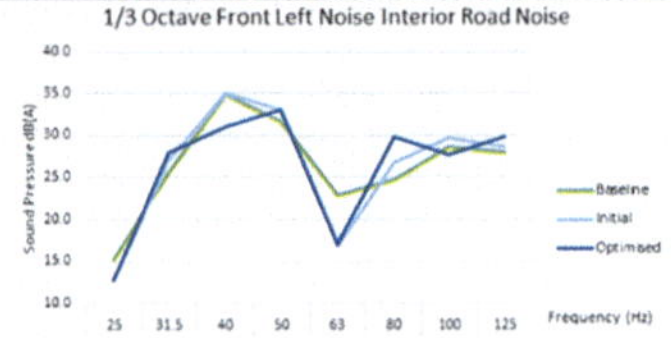					
Battery Pack	Front Crush	Force	kN	N/A	250.0		288.0		15%
		Intrusion	mm	N/A	2.6		3.8		44%
	Side Crush	Force	kN	N/A	225.0		200.6		-11%
		Intrusion	mm	N/A	2.3		8.6		273%
	Rear Crush	Force	kN	N/A	205.0		203.4		-1%
		Intrusion	mm	N/A	10.0		8.4		-16%

Fig. 11. Engineering interpretation summary data

an equivalent steel battery pack would be 60 kg heavier than the aluminum version. Results indicated that stiffening the interface between the battery pack and the rear section of the vehicle (including a D-pillar 'hoop' structure) may mitigate the reduction in first bending frequency, however this was outside the scope of the study.

The process has worked well to optimize the mass of the battery integration with a comparatively small range of geometric variables across static and dynamic vehicle attributes. This method could be expanded to other vehicle features, for example the structural performance of modules within the pack and other body in white variables. Based on the experience of this study possible improvements to the process could include: more extensive pre-studies on the response throughout the design space, running the optimization multiple times to gain confidence the optimum is found, coupling of a CAD package with an FE meshing tool to give greater freedom in parameterizing geometry, use of trimmed BIW modal analysis to make the process more representative, and inclusion of manufacturing constraints in the optimization.

4 Development of Sporty Sound vs. Cruising Comfort and Environmental Noise Pollution

Emotional engagement, driver feedback and brand image have always been key attributes in vehicle design, particularly for sportier cars. Conveying these attributes, the voice of the car was its engine sound, telling the driver and onlookers of its characteristics and behavior. Electric vehicles are becoming increasingly popular – as well as air pollution benefits, these cars provide strong and responsive performance, but they have lost their voice. While this gives advantages for cruising comfort and environmental noise pollution, the communicative ability of engine sound has been lost. Electric cars struggle to elicit emotional excitement or an 'involved' driving experience – and this leads to weaker brand appeal and loyalty. Furthermore, pedestrians, particularly those with visual impairment, are potentially endangered.

Electronic power unit sound augmentation systems can help to alleviate these problems by delivering carefully-controlled power unit sound, replayed through speakers or other actuators inside the cabin and outside the car. They provide an opportunity to tangibly benefit the dynamic driving experience and thus the brand image – greater appeal leads to faster take-up of electric cars, bringing forward air quality benefits. Electronic sound systems also improve pedestrian safety by informing bystanders about the current and imminent behavior of a car, thus reducing the risk of collisions.

However, there are risks – augmented sound can appear 'fake' (in most cases detracting from brand appeal) and it also has the potential to generate unwanted noise pollution.

Information content through power unit sound is key to delivering both a dynamic driving experience inside the car and pedestrian safety outside the car. When the driver is demanding acceleration by depressing the accelerator pedal, a corresponding increase in sound level is informative and expected (see Fig. 12). And as the vehicle

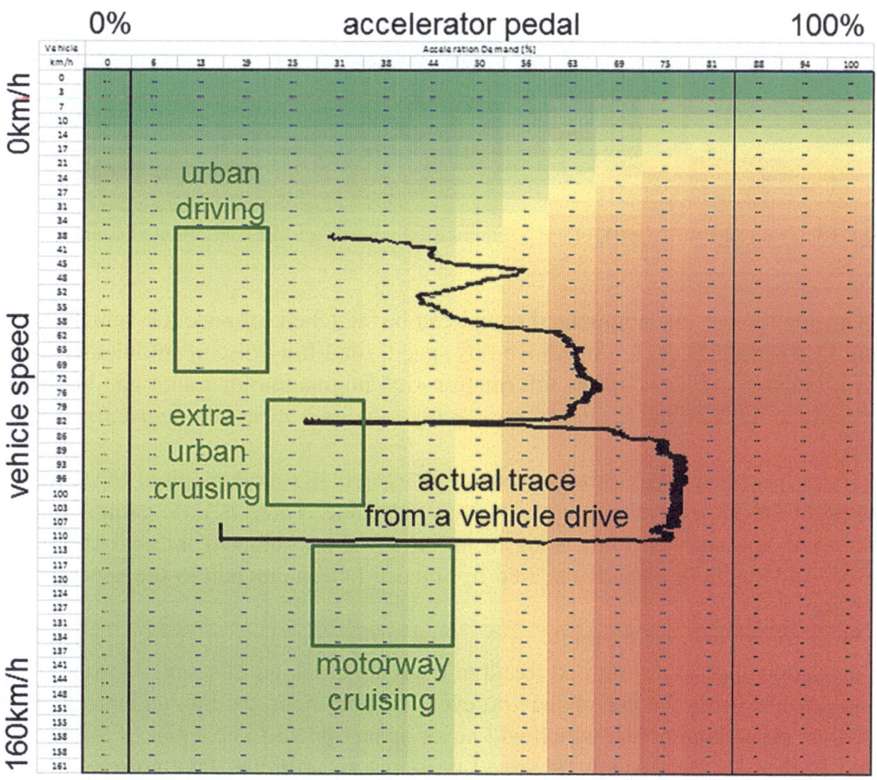

Fig. 12. Augmented sound gain map

travels faster, a corresponding increase in the sound's 'pitch' (frequency content) is informative. These two characteristics are constituents of the understood 'language' of engine sound, and are thus easy to understand, both for the driver and passengers and for bystanders outside the car.

Using the approach of maximizing the information content in the sound, the authors have learnt that the level of augmented sound can be minimized. This supports acoustic comfort in the cabin and lessens environmental noise pollution in the urban soundscape.

Under cruising conditions (with light accelerator pedal depression) the augmented sound's level can drop to near zero. The driver and passengers experience only the natural wind and road noise present inside the car (see also Sect. 3) at the lowest possible sound level, giving an acoustically comfortable cruise. With greater accelerator pedal depression, sound can match vehicle performance so that the acceleration feels exciting (and the driver knows that the battery is depleting). These changes in the augmentation sound can be seen in Fig. 13. Parallel calibration of accelerator pedal and augmented sound yields the most optimum result.

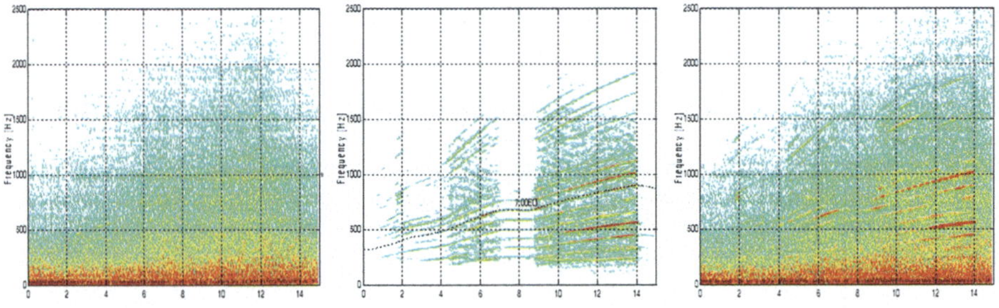

Fig. 13. Natural sound (left), tuned augmentation sound (middle), augmented sound (right)

Outside the car, the augmented sound can be switched off once the vehicle is travelling faster than 20–30 km/h. Above this speed, road-tire noise is sufficiently loud to inform pedestrians. Low-speed information-rich augmentation sound can be set at a level only audible to those near to the car (as long as it meets the legislative requirements).

There has been great debate over the character of sound required to convey the 'power' of an electric vehicle without appearing fake. There is little natural sound to build upon. Consumers don't want an imitation of an outdated internal combustion engine. And synthetic sounds inspired by science fiction can add to the appearance of fakeness.

Most sound systems currently in the market use synthesis to generate sound, giving the sound engineer unlimited freedom to create any kind of sound. However, if the sound deviates from the car's behavior, even for a moment, the human brain identifies a separate perceptual object which will never again appear to be a part of the car. Separated from reality, the synthetic sound becomes an artificial, fake noise, a situation which is particularly unacceptable in the purist sports car market.

The system named Realistic Augmented Sound by Ricardo (RAS-R) [9, 11] avoids synthesis. Instead, a sensor on the power unit provides a live-streamed sound signal that carries authentic and genuine information. This authenticity is intrinsic – the sensor 'measures' the car's actual physical behavior and cannot deviate from it. The engineer consequently has less freedom for sound design, because the sound type cannot fundamentally be changed. On the other hand, the signal contains genuine natural changes in level depending on load, and changes in frequency depending on speed. A toolbox of strong filters can further shape the sensor signal and produce a pleasing variety of sonic timbres, tuneable to meet brand image requirements. In particular, the load-tracking resonant low-pass filter is a highly effective tool to enhance accelerator pedal dependency. The authors' experience is that the information content and its authenticity carry through the filtering very effectively. The authenticity so delivered has been found to be very beneficial – both for driver engagement and to inform pedestrians of the current and imminent behavior of the vehicle and thus the nature of the hazard.

A recent development has been to combine both authentic sensor-based sound and synthetic sound in a multi-layered ensemble, leveraging the advantages of both approaches. The feeling of authenticity and immediacy of information come from the sensor-based sound. The synthetic sound can reinforce the sound when necessary. For example, the American AVAS standard includes a requirement for sound at zero vehicle speed – a notification to pedestrians that the car is ready to move. But at zero speed, the e-machine does not generate any vibration. A synthetic or pre-recorded, looped signal is needed. Another example is the use of synthetic low-frequency tones to strengthen the naturally higher-frequency bias of the sensor-based sound. This approach broadens the range of sound design possibilities, while retaining genuine information content in the augmentation sound.

5 Conclusion

NVH teams in automotive organizations have become adept at balancing interior and exterior NVH requirements with other vehicle attributes such as efficiency, dynamics, comfort and performance feel, as well as of course weight and cost. The efficient achievement of this balancing requires the following:

- Co-operation between different functional teams to tune the optimum attribute balance relating to specific market demands
- The ability to make detailed evaluations of NVH attributes alongside other attributes, in order to inform balancing decisions early in the product development cycle when design changes remain relatively cheap and easy

This paper has described three complex case studies exemplifying such trade-off situations, and the technical approaches taken to decide the best attribute balance for each vehicle product.

- Outstanding gasoline engine efficiency without excessive combustion noise

- Introduction of EV battery pack without detriment to ride comfort, driving dynamics or road noise
- Informative power unit sound in electric cars, without discomfort in cruising nor environmental noise pollution

References

1. Achieving a Vehicle Level Sound Quality Target by a Cascade to System Level Noise and Vibration Targets, Toru Mori, Akira Takaoka, Matthew Maunder, Society of Automotive Engineers SAE 2005-01-2394
2. March J., Strong G., Gregory S., Rediers B., Achieving Diesel Vehicle Appeal Part 1: Vehicle NVH Perspective, SAE 2005-01-2484
3. AEA website https://www.acea.be/statistics/tag/category/share-of-diesel-in-new-passenger-cars, last accessed 2019/05/21
4. ICCT Briefing.: CO_2 emissions from new passenger cars in the EU: Car manufacturers' performance in 2017. July 2018
5. Lee B., Oh H., Han S., Woo S., Son J..: Development of High Efficiency Gasoline Engine with Thermal Efficiency over 42%, SAE 2017-01-2229
6. Carlson C., Kalsule D.: Combustion Mechanical Breakdown: A Comparison of the Multiple Regression Method versus the Coherence Method for a HSDI Diesel Powertrain, SAE 2011-26-0035
7. Broatch A., Novella R., Gómez-Soriano J., Numerical Methodology for Optimization of Compression-Ignited Engines Considering Combustion Noise Control, SAE 2018-01-0193
8. A2Mac1 Auto Reverse Tesla Model X P90D 2016 https://www.a2mac1.com/AutoReverse/default.asp?c=1&p=2#/product/722, last accessed 2019/05/21
9. AQSIQ and Standardization Administration of China "Lithium-Ion Traction Battery Pack and System for Electric Vehicles – Part 3: Safety Requirements and Test Methods", GB/T 31467.3-2015.
10. Maunder, M., Grant, P., and Mawdsley, D., "System for Realistic Augmentation of Sporty Engine Sound Quality," SAE 2016-01-1781
11. Maunder, M., "Experiences Tuning an Augmented Power Unit Sound System for Both Interior and Exterior of an Electric Car," SAE 2018-01-1489

Using 3-Dimensional Transfer Path Analysis to Predict the In-vehicle Electrical Power Steering Noise Performance from Wheel-to-Wheel Rig Measurements

Charbel Massoud(✉) and Yuriy Kandinov

ZF Technical Center, Washington Township, MI 48094, USA
charbel.massoud@zf.com

Abstract. During the development of an Electric Power Steering system (EPS), it is critical that NVH is incorporated early in the development to mitigate noises that the end customer will find objectionable (roughness, whine … etc.) before start of production. This includes the determination of critical noise rating drivers and propagation of noise to the driver's ear through several vehicle tests. However, early in development, vehicles are not easily available, making it difficult to identify significant noise concerns. Because of this, it is critical that predictive methods are used. This can be done with a proper vehicle acoustic model being provided either by the OEM or by Transfer Path Analysis (TPA) conducted on a representative vehicle.

TPA has been used heavily throughout the automotive industry as a method for understanding the propagation of noises from the noise source to the driver's ear for several different automotive systems. However, conducting a TPA on an EPS at the vehicle level presents a challenge in visualizing and interpreting the data since a continuous speed sweep is not possible to cover the entire operating speed range of the EPS. Typically, several lock-to-lock iterations at different operating speeds need to be conducted to cover all operating conditions and results in several 2-dimensional plots to present and interpret. The approach of using 3-Dimensional Transfer Path Analysis improves the presentation and interpretation of Transfer Path Analysis for EPS Systems in the following ways:

- Reduce the analysis time effort by nearly 90 %
- Improve the resolution of the plots from 15 rpm to 0.5 RPM
- Allow for clear identification and tracking of orders which was not possible before
- Allows for in-vehicle SPL prediction in the same format as the standard in-vehicle test

W. Siebenpfeiffer (Hrsg.): *Automotive Acoustics Conference 2019,* Proceedings, S. 99–108, 2020. https://doi.org/10.1007/978-3-658-27669-0_8

Once a verified model is complete, it can be demonstrated that with force measurements from the ZF Wheel to Wheel Rig, where forces are measured directly, the in-vehicle noise performance of an EPS system can be predicted.

This approach significantly reduces the dependency on a vehicle by making it possible to identify noise risks from the simplicity of a bench test.

Keywords: Transfer Path Analysis · Vehicle prediction

1 Introduction

In vehicle NVH Development, it has been ideal to predict the noise performance of key systems/components early in the developmental phase, without the use of the vehicle, to optimize the system for noise prior to it ever being installed in a vehicle. This typically requires that a model of the vehicle or transfer paths are used to calculate the noise performance based on the input. The model can either be acoustic transfer functions of the vehicle calculated from CAE or acquired in Transfer Path Analysis. The input into the model should represent all possible operating conditions of the system to provide the most comprehensive understanding of the noise performance.

In rack drive electric power steering systems (EPS-RD), being able to represent all operating conditions (i.e. Hand Wheel Speeds and Rotational Direction) is difficult since acquisition time is limited by the travel of the rack (typically 2.5–3 handwheel revolutions. This makes it nearly impossible to plot in the speed domain. Also, being able to maintain a constant operational speed is also difficult providing a challenge to clearly distinguish the difference between orders and broadband noises in 2D plots at each hand wheel speed when plotting in frequency domain. The case can be made that to look at individual orders, one should calculate their plots in order domain as opposed to frequency domain. The addition of order domain plots increases the level of complexity in data reporting and can potentially lead to misinterpretation of the results.

In addition to the complexity defined above of displaying the results, conducting a Transfer Path Analysis (TPA), where the path contributions are displayed along with driver's ear measurements, can generate many plots that can be overwhelming to anyone trying to interpret the TPA. For example, a typical rack drive steering system with 4 mounts and measured at a variety of hand wheel speeds (typically 6 speeds at 15 rpm increments) can create up to 144 plots of each metric (e.g. Foot Mount Forces, Path Contributions … etc.) that will have to be reviewed to come up with conclusions on next steps for development.

$$4\ (\#\ \text{of mounts}) \times 3\ (\text{linear D.O.F.}) \times 6\ (\text{Speeds}) \times 2\ (\text{CW \& CCW}) = 144\ \text{plots}$$

A typical steering TPA report page is shown in Fig. 1. With 3 degrees of freedom on one page, the total number of pages can be simplified to 48. This can still make the interpretation of the TPA results an overwhelming task.

This paper will show that using 3D-TPA will reduce the complexity of the reporting of a TPA by reducing the number of report pages. In addition to the significant reduction of pages, 3D-TPA allows for the clear identification of orders and broadband noises. This paper will also show that using 3D-TPA makes it possible to comprehensively predict vehicle level noise performance from bench level tests using data from a case study that compares 2 gears (baseline and optimized) from the same program.

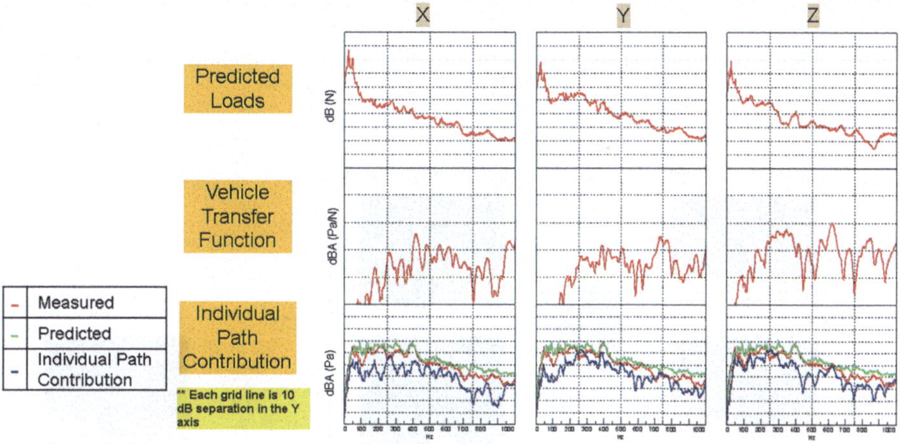

Fig. 1. Typical Steering TPA report page representing one foot mount at one handwheel speed in one handwheel direction. The top row shows the calculated foot mount forces for a specific path determined from TPA. The second row shows the acoustic transfer functions for the respective path. The bottom row shows the measured and predicted driver's ear noise performance as well as the respective path contribution. The XYZ coordinate system is based on SAE standard vehicle coordinates.

2 Advantages of 3D-TPA

3D-TPA incorporates a ZF proprietary tool called SOFA, which simplifies the presentation of steering data by presenting it into a 3-Dimensional plot. Figure 2 shows a typical 3D-TPA report page. The figure shows a TPA breakdown of the degrees of freedom on one mount in the vehicle. The first column of the figure shows foot mount forces that are measured or calculated from a standard TPA. From these forces and the Acoustic Transfer Function, which are also acquired during a standard TPA, individual path contributions can be determined. Summing the individual path contributions will provide the predicted (or calculated) interior noise performance based on the structureborne noise from the steering gear.

Figure 2 also shows that the handwheel speed and direction can be combined into one plot. Therefore, only one page per mount will be needed for a TPA report. For example, the 4 mount steering system identified above will reduce the number of pages from 48 to just 4. This significantly speeds up the analysis, reporting and improves interpretation of the results without the need for additional plots to identify orders or other key noise characteristics.

As shown in Fig. 1, only a small snapshot of the operating conditions can be acquired based on a single demanded handwheel speed. Each snapshot provides us with a desired resolution of 15 rpm. However, the actual speed can vary depending on how well the operator can maintain the desired speed. The figure shows that orders and broadband noises cannot be clearly identified because of the inconsistent handwheel speed. 3D-TPA uses the disadvantage of inconsistent speeds as an advantage to fill in the gaps between the 15 rpm increments, resulting in a resolution of 0.5 rpm. This aids in identifying key sound quality concerns such as orders and broadband noises.

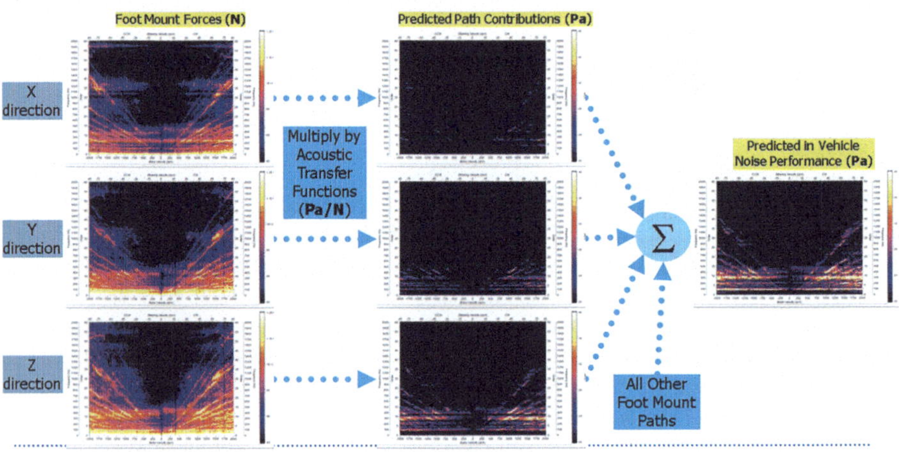

Fig. 2. A typical 3D-TPA presentation page representing one foot mount. The first column shows the Foot mount forces in 3 directions, 2nd column shows the respective path contribution and to the far right is the sum of all the path contributions (including other mounts) which is a prediction of the noise performance at the driver's ear.

As shown in Fig. 2, the orders and broadband noise can be clearly identified at all speeds. This contrasts with the difficulty of identifying these features in a 2D plot as shown in Fig. 1. From Fig. 2, contributions from the individual paths can be easily identified. For example, the predicted in vehicle noise performance shows lower frequency broadband noise and some order content. These characteristics are clearly seen in the Z-direction path contribution (2nd column, bottom plot). This is also clearly shown in the Z-direction of the foot mount forces (1st column, bottom plot). Another observation from Fig. 2 is that the force measurements (1st Column) show orders and broadband noises that don't necessarily appear in the path contributions (2nd column) because of the vehicle sensitivity to those particular paths. This can more clearly be identified in the 3D-TPA that is not observed in the 2D plots in Fig. 1.

One of the most important advantages to using 3D-TPA is simplifying how vehicle predictability based on Wheel to Wheel bench level testing can be verified. The main difficulty in doing this using Traditional TPA is due to maintaining the desired speeds defined above. These speed variations are not feasible to duplicate on a bench, since it would have to be done on a gear per gear basis. With the luxury of 3D-TPA, it is possible to take simple speed ramp profiles at the bench level and generate SOFA plots based on that input. With that, it is possible to utilize the block forces and predict the noise performance in the vehicle. The next section of this paper will discuss using 3D-TPA as a predictive method based on Wheel to Wheel bench testing.

3 Vehicle Prediction with 3D-TPA

As mentioned in the previous section, one of the main advantages to 3D-TPA is being able to verify predictability from block forces measured on the Wheel to Wheel rig. In this section we will look at the results from the bench testing on the Wheel to Wheel rig of 2 different gears from the same program and compare the results to the actual in vehicle measurements.

3.1 Specimen Description

To show the benefit of the EPS Vehicle predictive method, two gears from the same program were chosen to be evaluated at the bench level and vehicle level. One gear is a baseline design while the other is a gear with an optimized design. The changes are listed below:

- Improved motor design
- Improved Rack and Ball-Nut Assembly (RBNA) design
- Improved housing design for structure

3.2 ZF Wheel to Wheel Rig Results

The ZF Wheel to Wheel rig is a bench test used to measure the block forces of the steering gear foot mounts to characterize an EPS system's input into a vehicle[1]. This allows for the most accurate NVH measurement of a steering gear while being able to duplicate the boundary conditions of a vehicle.

For the sake of simplicity in this paper, only the 2 most critical paths will be plotted. Figure 3 shows a SOFA plot of the typical output of the block force measurements from the ZF Wheel to Wheel rig as measured on the baseline gear. Figure 4 shows the same mount measurements for the optimized design. Improvements from the optimized gear can already be observed in the block force measurements. However, it is

[1] The ZF Wheel to Wheel rig was featured in the January 2018 issue of ATZ Magazine, pp. 54–59.

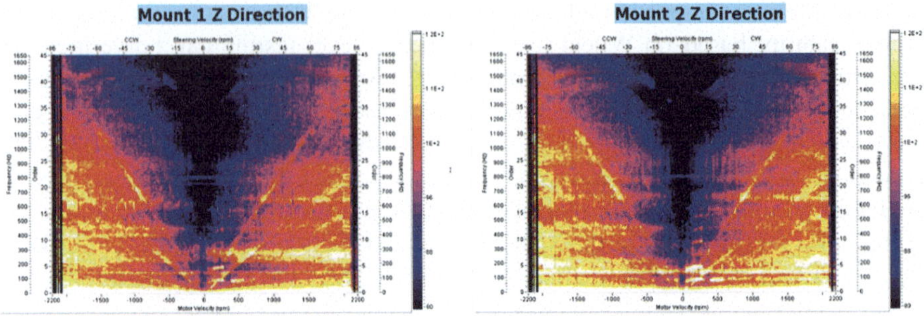

Fig. 3. Block force measurements of the baseline gear

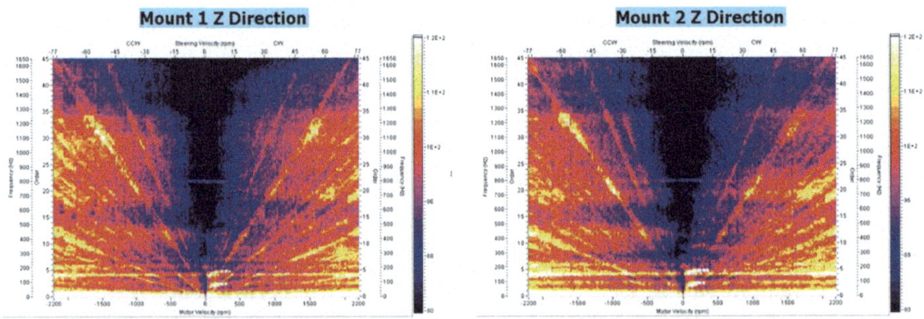

Fig. 4. Block force measurement of the optimized gear

important to show how much of an improvement can be found in the vehicle. Vehicle predictability will help us determine the level of improvement while showing other potential areas for improvement.

The acoustic transfer functions for the critical paths are shown in Fig. 5. This is necessary for the predictability of the in-vehicle NVH performance. Combining the acoustic transfer function with the block force measurements from the Wheel to Wheel rig gives us an estimation of the NVH performance in the vehicle. The same acoustic transfer functions were used for both the baseline gear and the optimized gear to show how the predicted in vehicle performance would look if both gears were installed in the same vehicle.

Figure 6 shows the predicted in-vehicle NVH performance of the baseline gear defined above. The key noise concerns identified are multiple broadband noises found between 200 and 1000 Hz. In addition to the broadband noises, a 45th motor order was observed.

Figure 7 shows the predicted in-vehicle NVH performance of the optimized gear. A significant reduction in the broadband noises is observed as well as a reduction in

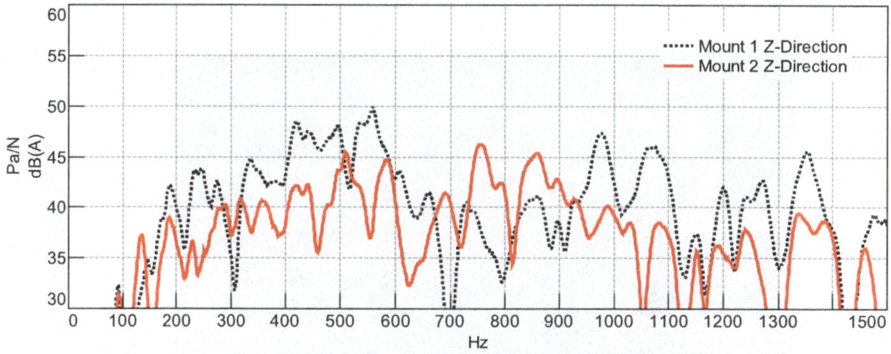

Fig. 5. Acoustic Transfer Functions from the critical paths presented to the driver's ear

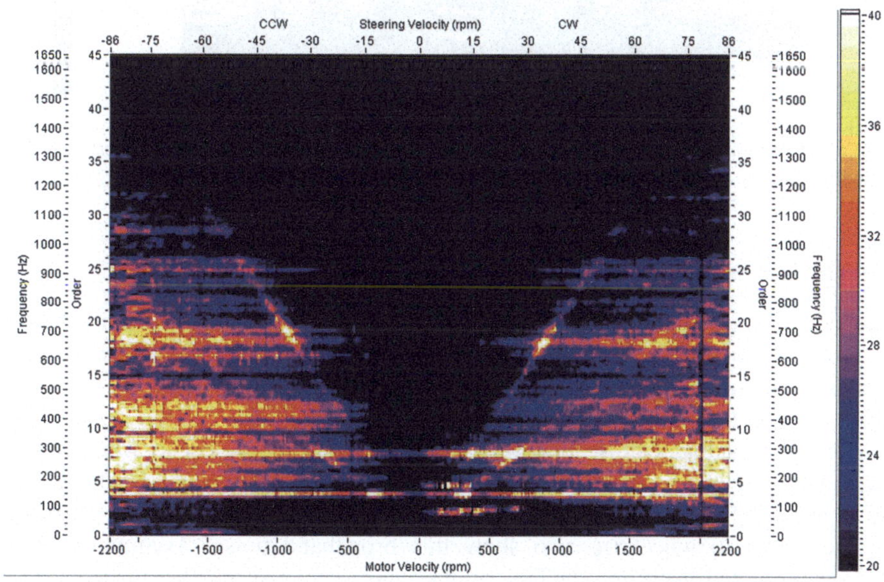

Fig. 6. Predicted vehicle noise performance of a baseline EPS steering gear based on Wheel to Wheel rig measurements

the 45th order. However, an increase in the 45th order above 1000 Hz is observed due to a structural resonance introduced there from the changed housing design. In the next section, the predicted in vehicle noise results will be compared to actual in vehicle measurements. The next section will highlight if the predictability of the in-vehicle measurements from the Wheel to Wheel rig can identify the improvements observed at the vehicle level.

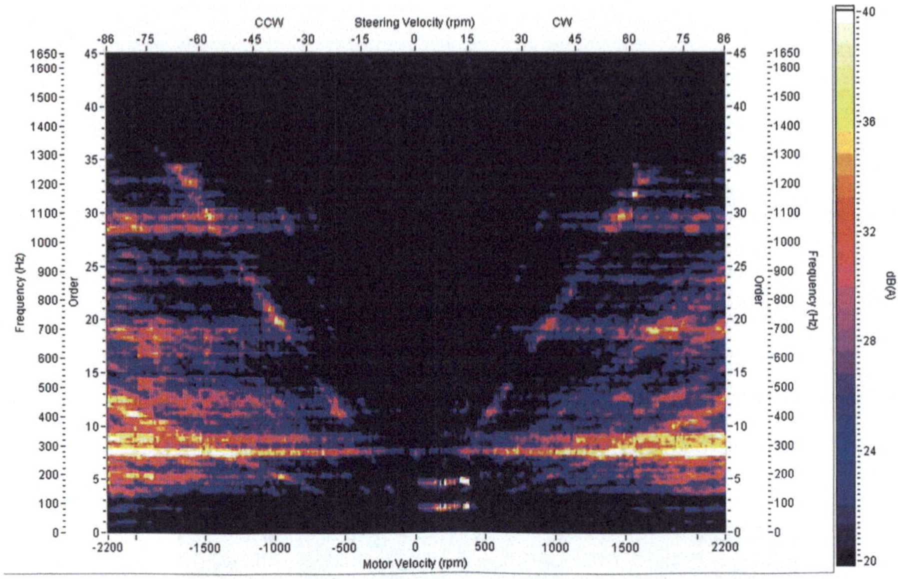

Fig. 7. Predicted vehicle noise performance of an optimized EPS steering gear based on Wheel to Wheel rig measurements

3.3 Vehicle Predictability

In the previous section we established a prediction of the in-vehicle noise performance. In this section we will compare the prediction to actual in-vehicle measurements using the same 2 gears.

The in-vehicle measurement results with the baseline gear is shown in Fig. 8. Much of the broadband noise that was predicted from the wheel to wheel rig (as shown in Fig. 6) is seen in the vehicle measurements. For example, in the vehicle there is noticeable broadband noise between 200 and 500 Hz. The predicted results from the wheel to wheel rig also show this broadband noise. Asymmetry above 500 Hz is predicted from the 3D-TPA on the wheel to wheel rig that is also seen in vehicle.

Although several similarities between the noise performance predicted from wheel to wheel rig and the in-vehicle measurements can be observed, some differences between the both methods can also be seen. Orders that were predicted with 3D-TPA from the wheel to wheel rig had been overestimated from what was observed in the vehicle (i.e. 45th order). Low frequency broadband levels were also overestimated with 3D-TPA versus what was measured in the vehicle. Despite the overestimations from the bench test, the noise concerns that were identified on the wheel to wheel rig were observable in the vehicle.

Figure 9 shows the vehicle measurement with the optimized gear. The figure shows the improvements that were anticipated from the 3D-TPA results from the

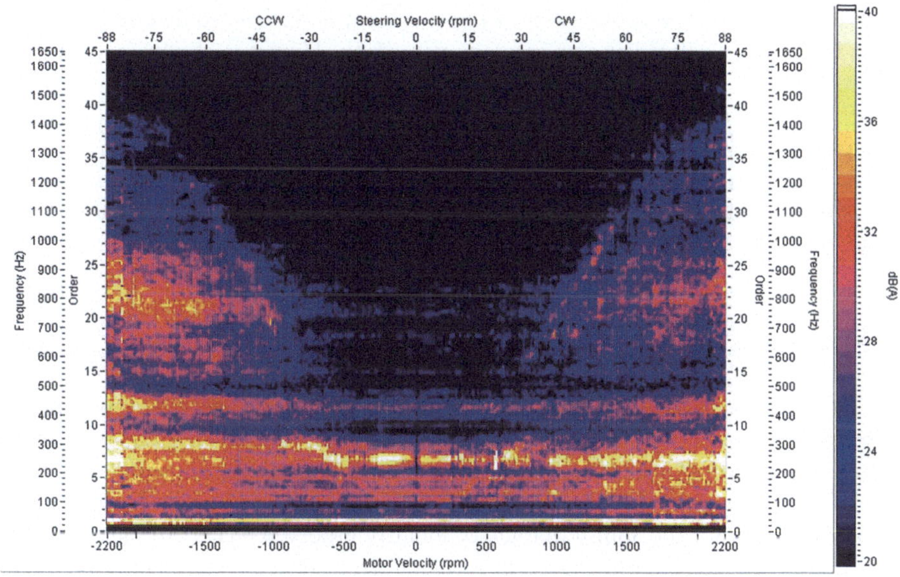

Fig. 8. In vehicle measured results of the baseline EPS steering gear

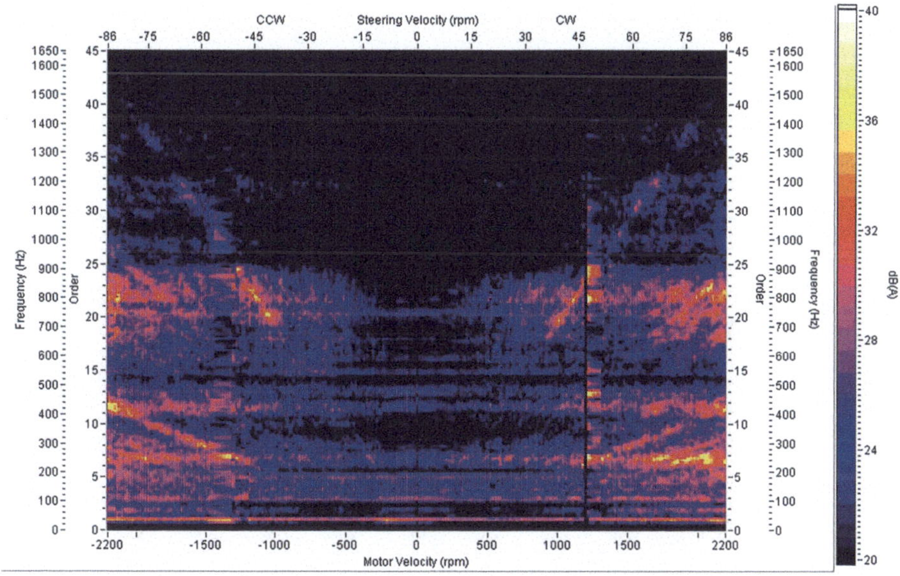

Fig. 9. In vehicle measured results of the optimized EPS steering gear

wheel to wheel rig. The actual level of improvement is difficult to determine because of the differences defined in the previous paragraph. However, directional improvement was noted. Also, the structural resonance above 1000 Hz that was observed at the bench level can be seen. However, it is not a significant influence on the overall noise performance.

The overall subjective rating improvement in the vehicle from the baseline gear to the optimized gear was nearly 1.5–2 points thus showing that critical NVH concerns can be identified and mitigated using 3D-TPA and the wheel to wheel rig.

4 Summary

The 3D-TPA has shown that it can significantly simplify the analysis, reporting and interpretation of a TPA conducted on an EPS system through the reduction of the number of plots/pages needed to present the TPA. 3D-TPA also improves the resolution of the plots, thus making the visualization of the plots much more comprehensive and helps to improve the ability to track or identify orders 3D-TPA also allows for the prediction of the noise performance in a vehicle from bench level block force measurements. A case study of the method shows that the vehicle noise concerns, although not perfectly, can be predicted from the block forces measured on the Wheel to Wheel rig. As well, improvements can be made at the bench level and predicted using the same method without the need for a vehicle.

Target Setting for Vibration Transmission Through Driveline Components Based on On-Vehicle and On-Bench Evaluation

Bart Bergen(✉), Jaysing Ashwin Chavan, and Kris Van De Rostyne

Toyota Motor Europe, Zaventem, Belgium
bart.bergen@toyota-europe.com

Abstract. The different connections between the powerplant and vehicle chassis/body should be developed carefully in order to minimise the transfer of unwanted vibrations. The engine mounts are well known to have significant contribution in the total transfer of powertrain noises in a broad frequency band, ranging from low-frequency booming noise to high-frequency gear noises. Also the driveshafts can often be an important contributor.

In order to control the NVH performance during the development of those components, clear component KPI's and targets are needed. A suitable KPI should fit in a consistent breakdown of the performance, e.g. based on a (simple) model of the considered contribution path. It is shown that an output/input side transmissibility can be a suitable KPI for driveshafts and engine mounting brackets.

In this work, the use of a transmissibility KPI is illustrated on two case studies: one for engine booming noise and one for gear noise. It is shown how the proposed KPI can fit in a robust target cascading process that can be used throughout the different project phases.

Since the final host vehicle is often not available during the component development, it is important that the selected KPI's can be evaluated on a test bench (or from an isolated component model). The correlation between on-vehicle and on-bench evaluation is reported, including an investigation of the trade-off between evaluation accuracy and bench complexity.

Keywords: Vehicle NVH · Powertrain · Target cascading

1 Introduction

The overall NVH performance of a vehicle depends on the individual performance of a large number of systems and components. The powerplant is typically one of the main noise and vibration sources, covering a broad range of frequencies and excitation phenomena (inertia and combustion forces, gearing excitation, excitation from

W. Siebenpfeiffer (Hrsg.): *Automotive Acoustics Conference 2019,* Proceedings, S. 109–121, 2020. https://doi.org/10.1007/978-3-658-27669-0_9

auxiliaries, ...). It is connected to the vehicle body mainly through the engine (or motor) mounting system, and through the driveshaft(s). A good NVH design of those systems is therefore critical to ensure a good isolation of the powerplant excitation and thus a good passenger comfort.

Given the large number of components involved, the management of the performance of those components should be organized in such a way as to make the definition of requirements and the follow-up of the performance during the development practically feasible [1–3]. This performance management is often organized following a 'V-model' (or V-shape). On the left leg of the V, performance requirements are cascaded down from full vehicle towards the different systems and components involved. This cascading is based on a suitable model of the performance. In the case of NVH, the first step of the cascade is typically to set contribution targets, and decide the input and body transfer target based on a Transfer Path Analysis [4–6].

In many cases, such an input or body transfer target is not directly useful for a component designer or supplier. Another breakdown step is needed to translate the body transfer or input targets into a specific component requirement. A key ingredient of this step is the definition (or selection) of suitable Key Performance Indicators (KPI's). A suitable KPI should fit in a consistent breakdown of the performance, e.g. based on a (simple) model of the considered contribution path. It is also important that the selected KPI can be evaluated with reasonable effort, either through CAE or physical measurements.

This paper presents two case studies on suitable KPI development for driveline components. A first case looks at the transmission of booming noise through the driveshaft (transmission of vibration of the powerplant through the vibration and bending of the driveshaft). In a second case, structure-borne gear noise is investigated. The focus is on finding a suitable KPI to characterize the performance of the engine mounting brackets. It is shown that an output/input vibration transmissibility can be a suitable KPI for both cases. It can fit in a consistent breakdown of the considered contribution, and is fairly straightforward to measure on-vehicle. Moreover, it has a clear physical interpretation. It is commonly used e.g. to describe vibration isolation of mounted systems. For a simple mass-spring system, the meaning is clear and straightforward. Also for actual components, the behaviour still reasonably resembles such a simple system in many cases.

Since the final host vehicle is often not available during the component development, it is important that the selected KPI's can be evaluated on a test bench (or from an isolated component model). It is clear that the boundary conditions on the test bench have a big impact on the evaluation of the proposed KPI's. One solution, then, is to design the test bench to represent as closely as possible the correct boundary conditions. This approach gives good results but is often challenging in practice as is illustrated through an application on the booming noise case. An alternative approach is to use substructuring theory [7, 8] to correct the evaluations for any difference in boundary conditions. The potential of this concept is illustrated on the gear noise case.

2 Target Cascading for Drivetrain Components

Good component NVH KPI's and targets are essential to achieve the full-vehicle performance at the end of the development [1–3]. Component targets are commonly obtained by cascading the full-vehicle targets following the 'V-model'. This is illustrated in Fig. 1, with a focus on the transfer systems: the different systems connecting the powerplant to the rest of the vehicle:

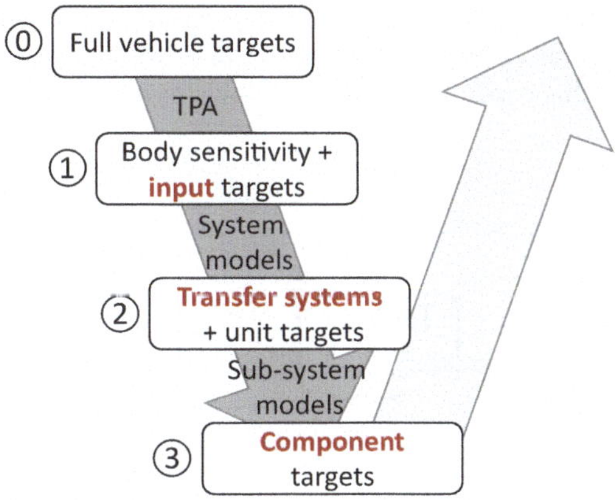

Fig. 1. Schematic illustration of the target cascading process, focused on the transfer systems between powerplant and vehicle

- The first step of the cascade is usually a Transfer Path Analysis (TPA) [4–6]. This method is well established, and benchmark data for body sensitivity and (to a lesser extent) input forces is readily available to support the first step of the target cascading.
- In a second step, this input force target is broken down further based on a (simple) model of the considered transfer system. At this level, the output/input acceleration transmissibility is proposed as KPI for the 2 cases considered. At the input side, the (free) vibration of the powerplant can reasonably be considered invariant.
- In the final step, the effect of the boundary conditions imposed by the assembly of the component on its performance in terms of the selected KPI is considered. Since the final vehicle is usually not available during the component development, the component boundary conditions have to be either recreated with sufficient accuracy or compensated through a suitable model to allow assessment of the component performance.

In the following sections, the second and third step of the breakdown are reported in more detail for two application cases: one on transmission of vibrations through the driveshaft, contributing to booming noise; and a second one on the transmission of gear noise through engine mounting brackets.

3 Case Study: Booming Noise Transmission Through Driveshaft

3.1 Problem Description

The first case study investigates the problem of engine excited noise. In addition to other structure-borne (EG mounts, exhaust hangers, …) and airborne (engine radiation, intake/exhaust nozzle noise, …) contributions, the driveshafts can contribute significantly to the interior booming noise. Engine vibrations excite the inboard joint through the transmission casing. This vibration is then transmitted through the driveshaft, and transferred to the knuckle, where it will excite the suspension and ultimately the body. This path is illustrated in Fig. 2. In this paper, the focus is on low-frequency noise (booming).

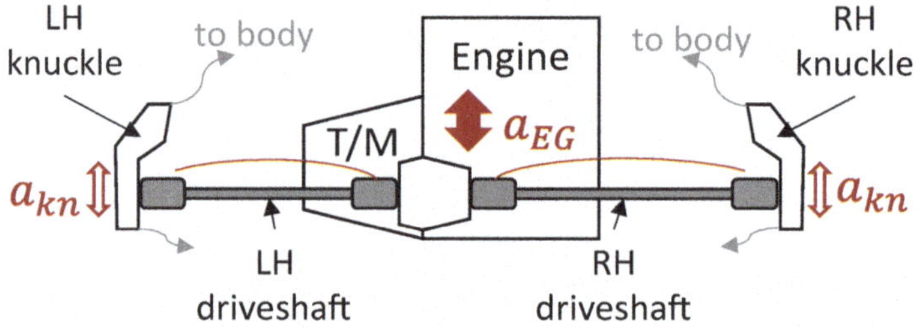

Fig. 2. Typical layout for a FF engine. Booming vibrations are transmitted through the driveshafts from the powerplant to the knuckles. (LH: Left-Hand, RH: Right-Hand, a_{EG}: engine acceleration, a_{kn}: knuckle acceleration)

3.2 Target Cascading for Driveshaft Booming Noise

KPI Selection and Breakdown Definition. At the first level of the cascade (refer to Fig. 1), the driveshaft contribution is composed of an input force and a body sensitivity (Fig. 3):

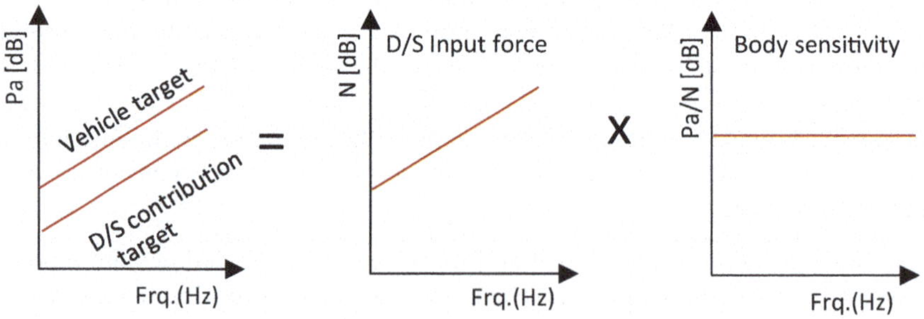

Fig. 3. Breakdown of driveshaft contribution in input force and body sensitivity (D/S: driveshaft)

The aim is to find a suitable description of the driveshaft input force, corresponding to step 2 of the cascade. Focussing on a driveshaft in Fig. 2, the inboard side is excited by the engine vibration a_{EG}; the outboard side is supported by the knuckle, characterized by an equivalent mass and stiffness M_{eq} and K_{eq} (Fig. 4):

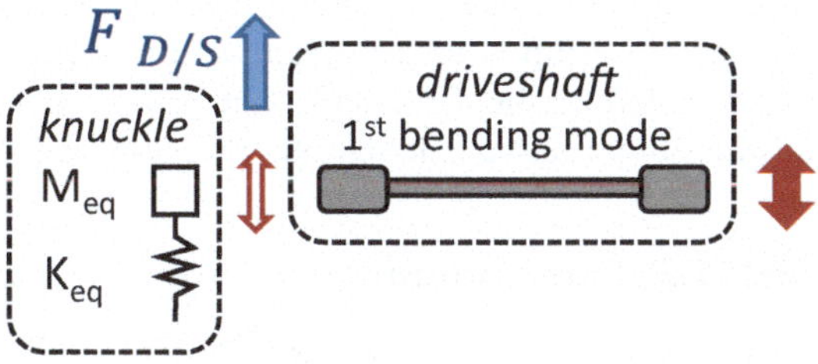

Fig. 4. Schematic view of the driveshaft and connections

As stated earlier, on the input side of the driveshaft, the unit acceleration a_{EG} can reasonably be considered invariant (if the considered frequency is sufficiently above the rigid body modes of the powertrain). On the other side, the driveshaft is connected to (supported by) the knuckle. If the outboard side acceleration a_{kn} is known, the force transferred to the knuckle can be calculated using the equivalent stiffness of the knuckle. It is clear from this image that the design of the driveshaft will impact the considered NV performance through the transmissibility ${}^{a_{kn}}/_{a_{EG}}$. Therefore, this transmissibility is proposed as KPI to characterize the driveshaft performance. Using this KPI, step 2 of the cascade can be made as:

$$F_{D/S} = a_{EG} \cdot \frac{a_{kn}}{a_{EG}} \cdot \left(\frac{a_{kn}}{F_{D/S}}\right)^{-1} \tag{1}$$

with $a_{kn}/F_{D/S}$ the knuckle intertance. This is illustrated in Fig. 5.

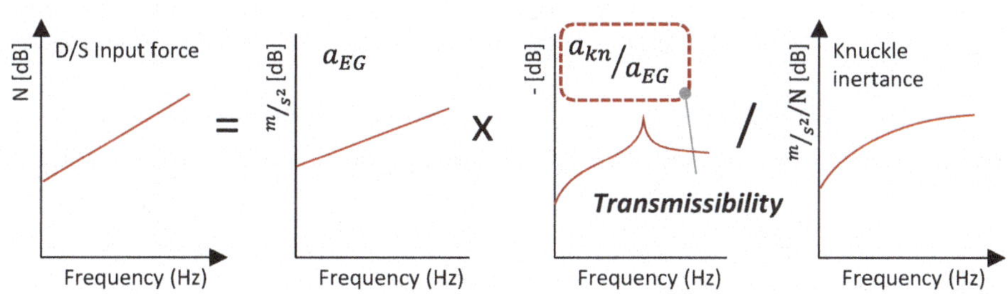

Fig. 5. Breakdown of driveshaft input force to unit input (a_{EG}), driveshaft transmissibility (${}^{a_{kn}}/_{a_{EG}}$) and knuckle inertance

Application Example. The proposed breakdown is illustrated on an application example. In Fig. 6, the different graphs of Fig. 5 are replaced with measured data from a passenger vehicle. The transmissibility is capturing well the impact of the driveshaft bending resonance, which can be seen in this example at 150 Hz. As such is easy to read and interpret. For reference, a force estimation from a matrix inversion TPA is added, showing the good accuracy obtained from the transmissibility calculation. This is also confirmed in a direct measurement of the interior noise, where the area around 150 Hz, dominated by the driveshaft contribution, is matching well with both the TPA and transmissibility based calculation. This is shown in Fig. 7.

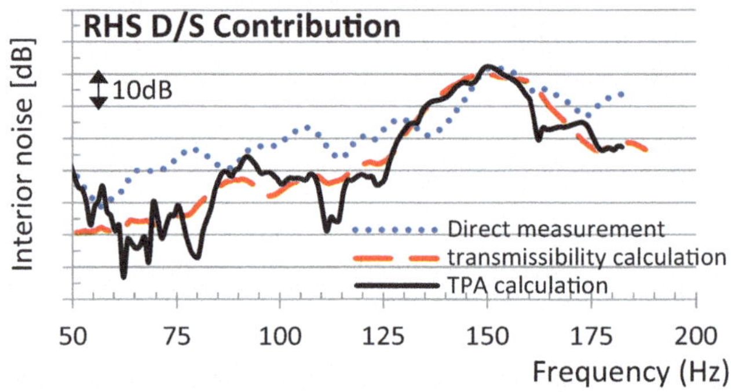

Fig. 6. Validation of the cascade for interior noise.

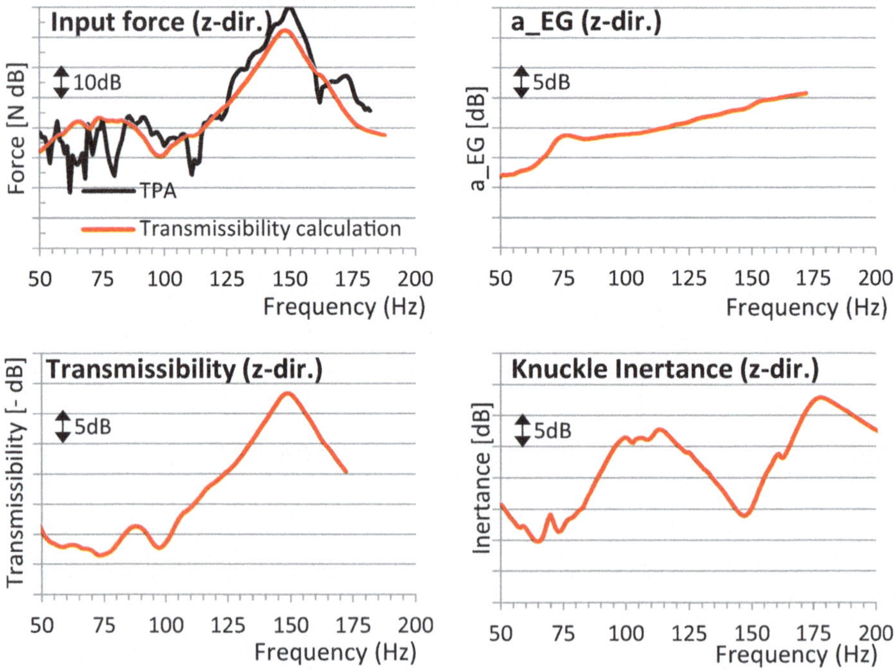

Fig. 7. Application example for the proposed breakdown of driveshaft input force

3.3 Performance Confirmation for the Driveshaft

The good accuracy of the performance breakdown based on driveshaft transmissibility shows the suitability of this KPI to characterize the driveshaft performance during its development. To this end, it is important that this transmissibility can be evaluated in an easy and accurate way, and – as much as possible – without the need for a prototype vehicle to mount the part.

Recalling the schematic of Fig. 4 and the definition of the transmissibility as $^{a_{kn}}/_{a_{EG}}$, two points can be noted:

1. The transmissibility can be evaluated from any measurement of a_{EG} and a_{kn} as long as a_{kn} is coherent with (caused by) a_{EG}. This can be an operational measurement (with a coherence filter on a_{kn} if necessary) or a hammering measurement with excitation on the powerplant side. The latter case does not strictly require the driveshaft to be mounted in a vehicle.
2. It is obvious that the boundary conditions at the knuckle side – specifically the equivalent stiffness (and mass) of the knuckle – will impact the transmissibility of the driveshaft. As such, the transmissibility is not a property of the driveshaft only, but will depend on the system in which it is assembled. In order to evaluate the transmissibility on a test bench (separate of the vehicle), this equivalent stiffness and mass will need to be recreated with sufficient accuracy. Alternatively, if a suitable model is existing, the impact of this boundary condition can be compensated through this model.

The first point is illustrated in Fig. 8, comparing the transmissibility obtained from an operational measurement and a hammering measurement (with the driveshaft installed on the vehicle). A reasonable agreement is observed; the mismatch is expected to be due at least in part to a different damping and stiffness of the joints between stationary and operating condition.

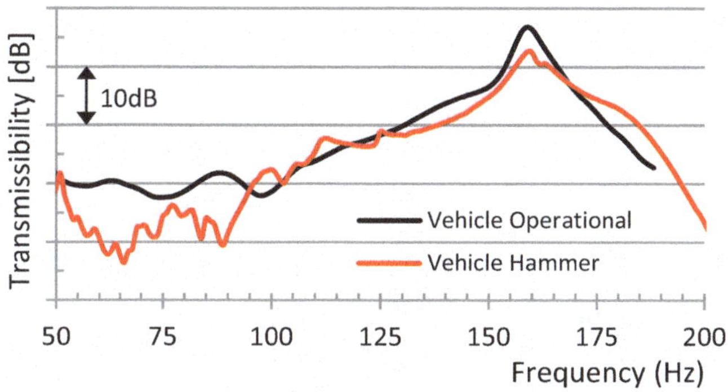

Fig. 8. Comparison of transmissibility evaluated from operational and hammering measurements (both on-vehicle).

As indicated in point 2, the stiffness at the outboard connection of the driveshaft will impact the evaluated transmissibility. In order to get the correct stiffness, the front suspension is rebuilt on a jig, including the strut, knuckle, and lower arm. The impact of including those parts is illustrated in Fig. 9. Finally, a good match between bench and vehicle evaluation can be achieved.

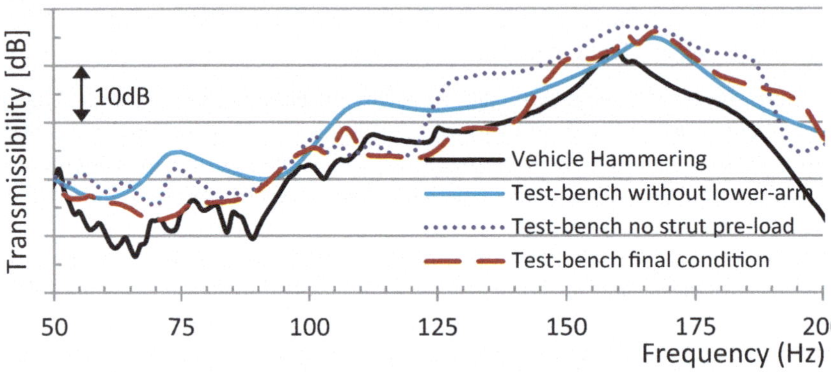

Fig. 9. Comparison of transmissibility evaluations by hammering: on-vehicle vs. on-bench, including effect of set-up conditions: strut pre-load and lower arm connection.

4 Case Study: Gear Noise Transmission Through Engine Mounts

4.1 Problem Description

The second case study considers the gear noise, specifically the structure-borne noise transmitted through the engine mounts. The discussion will focus on the transmission mount, but the conclusions should extend to other mounts as well. The problem is sketched in Fig. 10.

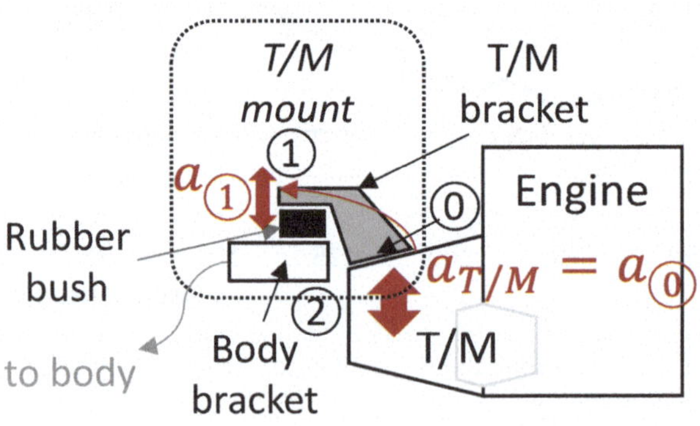

Fig. 10. Typical layout for a FF engine. Gear noise is transmitted through the transmission mount to the body. (T/M: transmission, $a_{T/M}$: transmission acceleration)

4.2 Target Cascading for Gear Noise Through the Engine Mounts

KPI Selection and Breakdown Definition. Similar to the first case study, the aim is to find a suitable description of the input force (or, by extension, the complete path), corresponding to the second step of the cascade of Fig. 1, for the mount contribution to gear noise. Considering the path in as illustrated in Fig. 10, a breakdown can be formulated as:

$$\frac{F_{②}}{a_{T/M}} = \frac{a_{①}}{a_{T/M}} \cdot \frac{F_{②}}{a_{①}} \tag{2}$$

The indices in $a_{⓪}$, $a_{①}$ refer to the location on the path, as indicated in Fig. 10. The ratio $F_{②}/a_{T/M}$ is basically the dynamic stiffness of the rubber bush. The performance of the transmission bracket is characterized by the transmissibility $a_{①}/a_{T/M}$, which is proposed as KPI for this case.

The connection from the bracket to the rubber can reasonably be assumed free, as the application for gear noise will typically be far above the powerplant rigid body modes. At the other side of the bracket (point ⓪), the stiffness of the T/M will not impact the breakdown as formulated above (its effect is considered part of the input acceleration $a_{T/M}$), but it will affect the practical evaluation of the transmissibility.

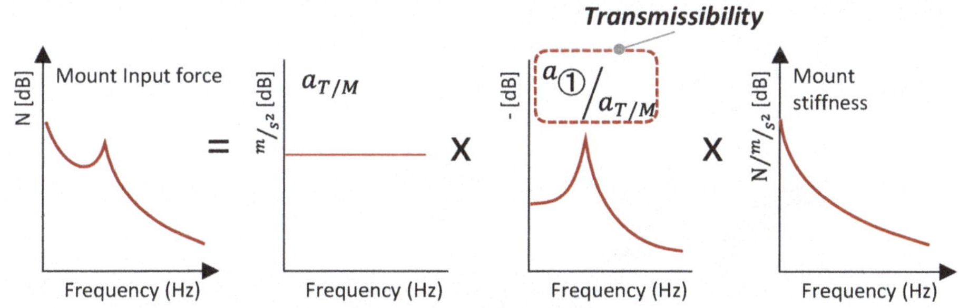

Fig. 11. Breakdown of the mount input force to unit input ($a_{T/M}$), T/M bracket transmissibility ($a_{①}/a_{T/M}$) and mount stiffness

This breakdown is illustrated in Fig. 11.

Application Example. Also for the second case, the proposed breakdown is illustrated through an application example for a passenger vehicle. Rather than a validation on the level of the mount input forces, the entire path is measured in the form of a transfer function $P/a_{T/M}$, with the T/M acceleration (at the root of the T/M bracket) as input and the interior noise as response. As such, a unit input is used instead of operational acceleration. For this example, a simple stiffness is used for the rubber bush. For final application, it would be advisable to use a detailed frequency-dependent stiffness for the rubber bush considered.

In Fig. 12, the transmissibility is plotted for the considered example. The behaviour on the actual part is very much in line with the expected response of a theoretical mass-spring transmissibility, making for an easy interpretation of the KPI. For this part, the first two resonances are very clear around 450 and 1300 Hz.

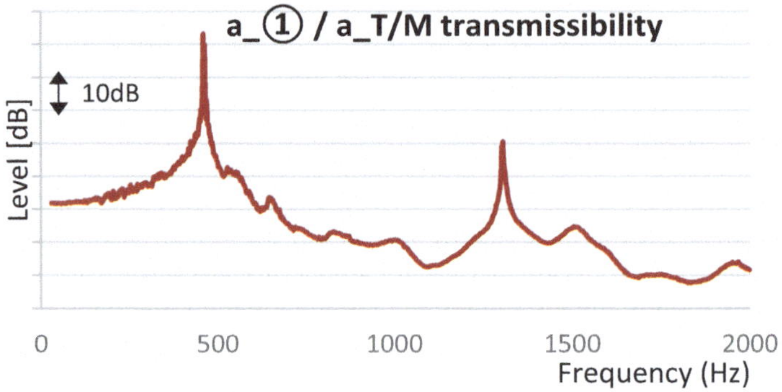

Fig. 12. Application example of the transmissibility $a_{\textcircled{1}}/a_{T/M}$

Finally, Fig. 13 compares the direct measurement of $P/a_{T/M}$ with a calculation of $F_{\textcircled{2}}/a_{T/M}$ according to Eq. 1, further multiplied with the corresponding body sensitivity from the body side bracket. In general, a good agreement can be observed, validating the proposed KPI. The impact of the bracket resonances (as a main property of the bracket) is well captured.

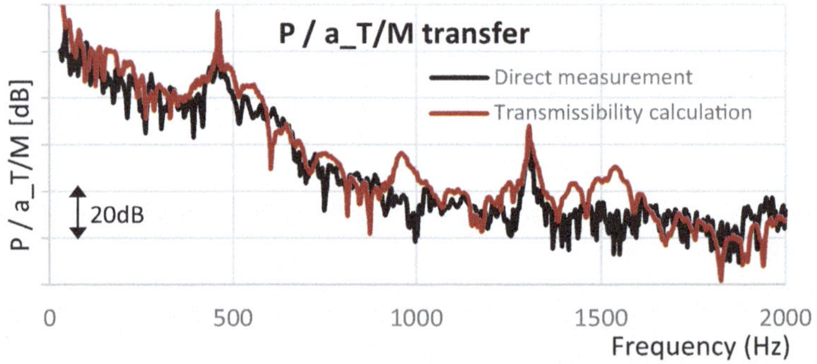

Fig. 13. Validation of the cascade for $P/_{a_{T/M}}$

4.3 Performance Confirmation for Engine Mount Brackets

The comments formulated above (Sect. 3.3) apply equally to this case. Again, the transmissibility can be evaluated both from operational data (ensuring coherent signals) or from a hammering measurement. Also for this case, the aim is to be able to evaluate that transmissibility without the need for the vehicle.

As indicated in Sect. 4.2, the main concern in this application case is the stiffness of transmission case at the bracket attachment point(s). As an alternative for relying the availability of a (prototype) transmission case, the impact of this connection stiffness could be introduced through a substructuring model. This allows the development and evaluation of the bracket as a standalone component.

Frequency-Based Substructuring. This method allows to recover the dynamic behaviour of a composite system based on FRF (frequency response) information of both systems referenced to the connection points [7, 8].

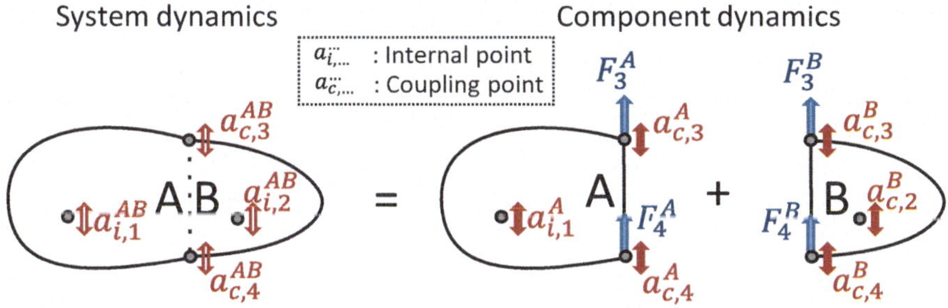

Fig. 14. Frequency-based substructuring concept for a system composed of two components A and B

For the illustration in Fig. 14, the dynamic response of the composite structure can be recovered according to:

$$H^{AB} = \begin{bmatrix} H_{ii}^A & H_{ic}^A & 0 \\ H_{ci}^A & H_{cc}^A & 0 \\ 0 & 0 & H_{ii}^B \end{bmatrix} - \begin{bmatrix} H_{ic}^A \\ H_{cc}^A \\ -H_{ic}^B \end{bmatrix} \left[H_{cc}^A + H_{cc}^B \right]^{-1} \begin{bmatrix} H_{ic}^A \\ H_{cc}^A \\ -H_{ic}^B \end{bmatrix}^T \tag{3}$$

Where the transfer matrices $H_{...}^{...}$ are defined as $F_i^A = H_{ii}^A a_i^A$. For this application, the substructuring method can be used to recover the transmissibility of the bracket assembled on the transmission casing based on a free-free measurement of the bracket ($a_{①}/a_{T/M}$ & $a_{T/M}/F_{T/M}$), and a measurement of the inertance ($a_{T/M}/F_{T/M}$) on the transmission case.

Application Trial – Proof of Concept. While the substructuring theory is certainly valid and applicable to this case, the procedure is quite sensitive to measurement errors (including mass loading) and noise due to the necessary inversion in Eq. 2 [9–11]. This issue is exacerbated by the often compact structure of a transmission bracket, and the limited accessibility to conduct the practical measurements.

These issues are reflected in the example of Fig. 15, showing the inertance at point ⓪, comparing a direct measurement of the assembly to a substructuring calculation. The spurious peaks on the response are due to conditioning issues, probably combined

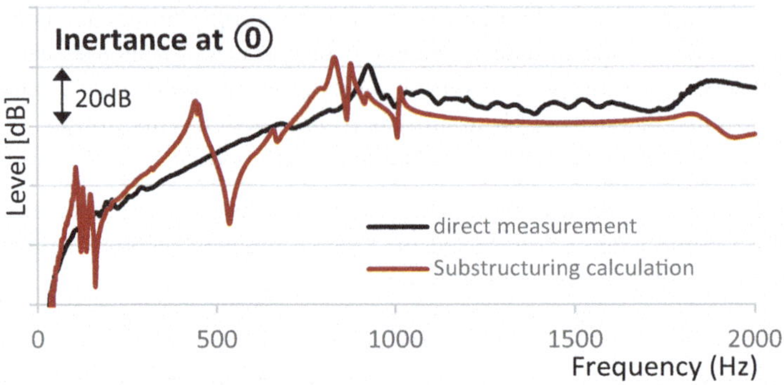

Fig. 15. Substructuring illustration for the inertance at point ⓪

with a mass loading error in the bracket measurements. Nonetheless, the general trend and level can be reproduced well using the substructuring method. Note that for this example, no condition control (e.g. through singular value decomposition) or FRF correction (e.g. modal filtering) was applied.

5 Summary and Conclusion

This paper proposes the use of a transmissibility KPI to characterize the dynamic performance of driveline components, specifically the driveshaft – when considered as a contribution to engine booming noise – and the engine mounting brackets – when considered for their contribution to gear noise. In two case studies, it is shown that the transmissibility fits well in a performance breakdown for those phenomena.

A general benefit of using transmissibility is the clear and easy interpretation of the quantity in terms of resonance frequencies and amplitude (mainly relevant in the decoupling area). The application examples show that components considered behave reasonably like a mass-spring system, further supporting this clear interpretation.

A good component KPI should allow a performance evaluation as a stand-alone part, independent of the vehicle. This is possible, provided that the interface of the considered component to the vehicle are suitably represented during the evaluation. One way to achieve this is by re-creating the dynamic behaviour (mainly stiffness) at the interface on a test jig. Alternatively, a substructuring method can be used to introduce the effect of the interface. Both concepts are illustrated in the case studies.

Acknowledgements. The authors gratefully acknowledge M. Komada from Toyota Motor Corporation, Japan, for his support of the activities reported here. Thanks also to Vincenzo Abbondante for supporting this work as a part of his internship.

References

1. Mori, T., Takaoka, A., and Maunder, M., "Achieving a Vehicle Level Sound Quality Target by a Cascade to System Level Noise and Vibration Targets," *SAE Technical Paper* 2005-01-2394 (2005), https://doi.org/10.4271/2005-01-2394.
2. Tousignant, T., Govindswamy, K., Tomazic, D., Eisele, G. et al., "NVH Target Cascading from Customer Interface to Vehicle Subsystems," *SAE Technical Paper* 2013-01-1980 (2013), https://doi.org/10.4271/2013-01-1980.
3. Bergeon, M., Braunwart, P., Daly, M., and Huber, J., "Vehicle Cascade & Target Response Analysis (VeCTRA) is an Excel Based Tool Used for the Idle NVH Target Cascade Process," *SAE Technical Paper* 2003-01-1434 (2003), https://doi.org/10.4271/2003-01-1434.
4. Van der Seijs, M.V., de Klerk, D. and Rixen, D.J., "General framework for transfer path analysis: History, theory and classification of techniques", *Mechanical Systems and Signal Processing* 68–69, 217–244 (2016), https://doi.org/10.1016/j.ymssp.2015.08.004.
5. Van der Auweraer, H., Mas, P., Dom, S., Vecchio, A. et al., "Transfer Path Analysis in the Critical Path of Vehicle Refinement: The Role of Fast, Hybrid and Operational Path Analysis," *SAE Technical Paper* 2007-01-2352 (2007), https://doi.org/10.4271/2007-01-2352.
6. Bergen, B., Schaefer, N., Van de Rostyne, K., and Keppens, T., "Vehicle Acoustic Performance Analysis towards Effective Sound Package Design in Mid-Frequency," *SAE Technical Paper* 2018-01-1495 (2018), https://doi.org/10.4271/2018-01-1495.
7. D'Ambrogio W., Fregolent A., "Direct decoupling of substructures using primal and dual formulation" In: *Proulx T. (eds) Linking Models and Experiments, Volume 2. Conference Proceedings of the Society for Experimental Mechanics Series. Springer, New York* (2011).
8. De Klerk, D., Rixen, D.J., Voormeeren, S.N., "General framework for dynamic substructuring: History, review, and classification of techniques" *AIAA Journal* 46 (5), 1169–1181 (2008) https://doi.org/10.2514/1-33274.
9. Nicgorski, D., Avitabile, P., "Conditioning of FRF measurements for use with frequency based substructuring" *Mechanical Systems and Signal Processing* 24(2), 340–351 (2010) https://doi.org/10.1016/j.ymssp.2009.07.013.
10. Nicgorski, D., Avitabile, P., "Experimental issues related to frequency response function measurements for frequency-based substructuring" *Mechanical Systems and Signal Processing* 24(5), 1324–1337 (2010) https://doi.org/10.1016/j.ymssp.2009.09.006.
11. Peeters, P., Manzato, S., Tamarozzi, T., Desmet, W. "Reducing the impact of measurement errors in FRF-based substructure decoupling using a modal model" *Mechanical Systems and Signal Processing* 99, 384–402 (2018) https://doi.org/10.1016/j.ymssp.2017.06.020.

Locally Resonant Metamaterial Patches on Rear Shock Towers of a Vehicle to Reduce Structure-Borne Road Noise: Numerical and Experimental Performance Validation

L. Sangiuliano[1,2](✉), C. Claeys[1,2], E. Deckers[1,2], J. De Smet[3], B. Pluymers[1,2], and W. Desmet[1,2]

[1] Department of Mechanical Engineering, KU Leuven, Leuven, Belgium
luca.sangiuliano@kuleuven.be
[2] DMMS lab, Flanders Make, Leuven, Belgium
[3] MotionS lab, Flanders Make, Lommel, Belgium

Abstract. Over the past decades, automotive companies have sought lightweight and performant noise, vibration and harshness (NVH) solutions to comply with stringent regulations for CO_2 emissions and noise pollution. Combining lightweight design with improved NVH solutions is often a challenging task: low mass and high stiffness materials are generally characterized by poor NVH behavior and low noise and vibration levels often require heavy and bulky additions, especially to be effective in the low frequency regime. To face this challenge, recently, locally resonant metamaterials (LRMs) have come to the fore. These materials combine in one solution lightweight design and superior noise and vibration attenuation performance, beating the mass law in tunable frequency ranges, referred to as stop bands. The LRM concept is used in this work to tackle a low frequency structure borne road noise problem in a commercial vehicle. A LRM solution is applied on the rear shock towers of the vehicle, with the goal of attenuating the vibrational energy entering into the vehicle body through the suspension assembly, which is excited by the interaction of the tire with the road while driving. This results in a reduction of the noise in the interior compartment around 190 Hz. In order to benchmark the performance of the LRM concept, a vehicle is chosen which is sold with a tuned vibration absorber installed on each of the rear shock towers as NVH solution. Each of the tuned vibration absorbers (TVAs) adds 1.46 kg of mass to the vehicle. The LRM concept is designed to reduce the mass of the current solution by 48% and to have similar NVH performance. The LRM concept is realized through additive manufacturing and it is added as patches on the rear shock towers to replace the TVAs. Both numerical and experimental results in lab and on a smooth road profile validate the performance of the LRM concept proposed.

Keywords: Locally resonant metamaterials · Vibrational energy attenuation · Structure borne road noise

W. Siebenpfeiffer (Hrsg.): *Automotive Acoustics Conference 2019*, Proceedings, S. 122–136, 2020. https://doi.org/10.1007/978-3-658-27669-0_10

1 Introduction

Over the past decades, automotive companies have been steering the production of their vehicles towards lightweight design, in order to reduce CO_2 emissions [1]. Parallel to this evolution, the demand for performant NVH solutions has increased due to the awareness of noise pollution on human health [2] and growing customer expectations. The use of lightweight materials often leads to unsatisfactory dynamic behavior, since the sound insulation that a partition can achieve scales with its own mass [3] and thus conflicting with the ever more stringent noise regulation. Novel vibro-acoustic solutions are then pursued that can satisfy the often conflicting requirements of lightweight design and performant NVH behavior.

Among these novel solutions, LRMs have come to the fore. These engineered materials exhibit stop band behavior. Stop bands are frequency ranges in which free wave propagation is impeded [4] and in which LRMs can achieve superior noise and vibration insulation properties, beating the mass law. In LRMs, stop band formation is due to Fano-type interference [5] and it can be achieved by local resonances due to discrete addition of resonant elements to an elastic host structure on a subwavelength scale [6–8]. Therefore, stop band behavior in LRMs does not require periodicity in the lattice and this makes them suitable for such applications (e.g. in vehicles) that demand compact and lightweight design with enhanced NVH properties in the-hard-to-address frequency range below 1 kHz.

In previous publications, the authors showed the potential of LRMs in different vibro-acoustic problems: sound transmission loss applications for partitions in different vibro-acoustic environments [9–11] and vibrational energy isolation applications for mechanical systems, consisting of one transfer path, such as a duct [12], and of multiple transfer paths, such as a simplified shock tower [13]. In this paper, a LRM solution is applied on the rear shock towers of the vehicle, with the goal of attenuating the vibrational energy entering into the vehicle body through the suspension assembly, which is excited by the interaction of the tire with the road while driving [14–20]. In order to benchmark the performance of the LRM concept, a vehicle is chosen which is sold with a TVA installed on each of the rear shock towers as NVH solution. These TVAs are designed to reduce the noise in the interior compartment around 190 Hz. The LRM concept is realized through additive manufacturing and it is added as patches on the rear shock towers to replace the tuned vibration absorber.

Both numerical and experimental investigations are presented. Numerical results calculated with a Finite Element (FE) model of the vehicle extend and validate the experimental finding reported by the authors in a previous publication [21].

This paper is organized as follows. In Sect. 2 the application case is introduced. In Sect. 3 the LRM solution design is defined. In Sects. 4 and 5 the FE model of the vehicle and the test setup are described, respectively. In Sect. 6, the numerical and experimental results are presented. Finally, in Sect. 7 the main findings of this work are summarized.

2 Problem Definition

In vehicles, vibrational energy generated by tire/road interaction while driving largely contributes to the low frequency structure borne noise in the interior compartment. This energy flows from the tire contact patch area into the vehicle body through the suspension assembly system, causing vehicle body parts vibrating and radiating noise [14, 15]. This vibro-acoustic problem typically occurs in a frequency band around 200 Hz and its dominant frequency depends on several mechanisms related to the dynamics of the tires and their interaction with the road profile, the suspension assembly systems, the driving conditions [14–20] …

The vehicle studied in this paper is a European SUV. It has MacPherson Strut rear suspensions and tires of type 235/60 R18. Because of the type of suspension assembly system and tire mode, a high noise level in the interior compartment is measured at 190 Hz. The SUV is equipped with a TVA on each of the rear shock towers (see Fig. 1). These TVAs are designed to be effective around 190 Hz, in order to reduce the vibrational energy entering the vehicle body through the rear suspensions and, thus, to reduce the noise level in the interior compartment. This current NVH solution adds 1.46 kg on each of the rear shock towers.

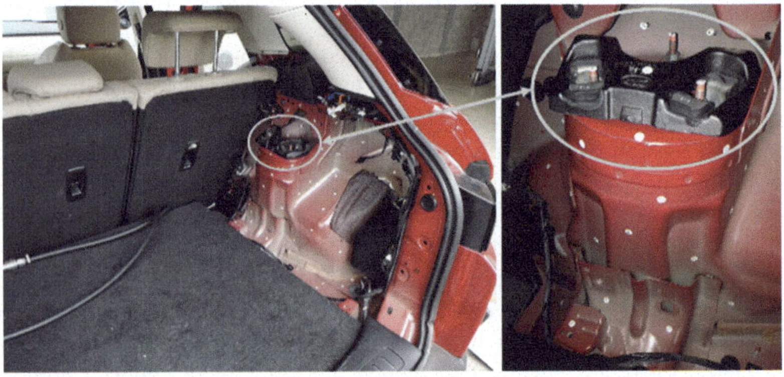

Fig. 1. TVA installed on the right rear shock tower of the vehicle [21].

The TVAs are replaced by patches composed by resonant elements. These patches are added onto a host structure, identified as the area where the shock absorbers are connected to the shock tower. The assembly composed by resonant elements and host structure makes up the LRM solution.

3 Design and Performance Prediction of the LRM Solution

The LRM solution is tuned to exhibit stop band behavior around 190 Hz, in the frequency range where the TVA is effective. In order to exhibit stop band behavior, the resonant element needs (i) to exert a non-zero net force onto the host structure [22], (ii) to introduce a localized resonant behavior in the host structure [9] and (iii) to be added in a sub-wavelength scale [24].

3.1 Design of the Resonant Element

In order to meet the first two conditions to exhibit stop band behavior, the design of the resonant element follows the same philosophy as previous work of the authors [9–13, 21], for more details, the interested reader is referred to those works. The resonant element is designed to resemble a mass-spring-damper system consisting of two cantilever beams with an end point mass on each end. A support connects the two beams and the base, the latter is the connection area between the resonant element and the vehicle (see Fig. 2 – left). The resonant elements are made of polyamide and they are realized by selective laser sintering.

The resonant element is designed using an FE model, consisting of 6807 linear solid CHEXA8 elements [25] and clamped base and verifying the numerical results experimentally on the printed resonant samples [21]. The geometrical dimensions of the resonant elements are updated until the first in-plane and out-of-plane bending mode of located at 193 Hz, which is close enough to the tuned frequency of the TVA (see Fig. 2 – right).

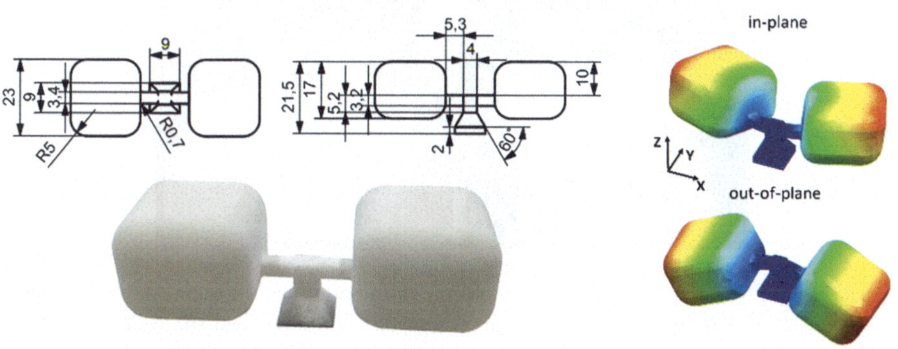

Fig. 2. (Left) Design and picture of the manufactured resonant element. The dimensions are given in mm. (Right) Mode shapes of the first bending modes of the resonant element at 193 Hz in in-plane and out-of-plane directions.

3.2 Position of the Resonant Elements on the Rear Shock Towers

The resonant elements are added onto the rear shock towers as patches. These patches are shaped in a similar form as the rear shock towers with the aid of a CAD model. To ease installation, four patches for each shock tower are made instead of one. The resonant elements in each patch are spaced in a sub-wavelength scale on a more or less regular grid of dimensions of 65 mm in x direction and 25 mm in y direction. A thin and flexible beam with rectangular cross section, 1 mm thick and 4 mm wide, connects the bases of the resonant elements.

The location and the minimum amount of resonant elements to be added on the shock towers, in order to prevent energy from propagating through the structure, are based on the findings in previous studies [13]. The resonant elements cover the area around the connections of the shock absorber to the shock tower. This area is expected to be part of the transfer path for the vibration energy coming from the tires and travelling into the vehicle body. The treated area extends to the surrounding vehicle body, in order to host a minimum amount of 5 rows of resonant elements, counted from the force input location to the farthest resonant element (Fig. 3).

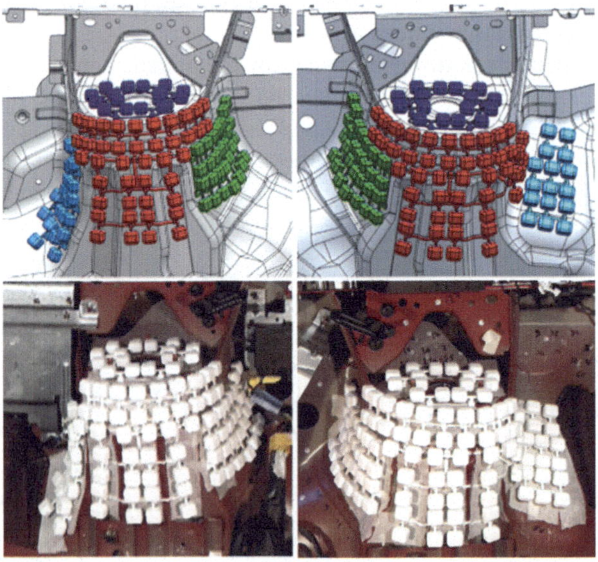

Fig. 3. (Top) CAD and (bottom) picture of the installation in the vehicle of the patches on the rear shock towers. Left and right figures represent the left and the right shock tower, respectively. The colors red, purple, green and blue in the CAD model indicate the four different patches [21].

The left and the right shock towers host 51 and 54 resonant elements, respectively, since the two shock towers are not symmetric. The patches are glued to the shock towers with superglue. In total the two patches combined weigh 1.52 kg, reducing the added mass as compared to the TVA by 48%.

3.3 Performance Prediction of the LRM Solution

Stop band behavior is the LRM system is predicted using Unit Cell (UC) modelling approach [24, 25]. A flat steel plate of 2 mm thickness is used to describe the UC of the shock tower. The material properties of the steel are obtained from previous work [21]. For modelling purpose, the resonant elements are added onto the host structure on a periodic grid of dimensions 65 × 25 mm, although stop band behavior in locally resonant metamaterials do not rely on periodicity [23]. These dimensions define the geometry of the UC of the LRM system analyzed (see Fig. 4). The UC is modeled with FE modeling approach and periodic Bloch-Floquet boundary conditions [26, 27]. The host structure is modeled with 429 linear shell CQUAD4 elements [25], whereas the resonant element is modeled as described in Sect. 3.1. For the dispersion curves results and discussion, the interested reader is referred to this work [21].

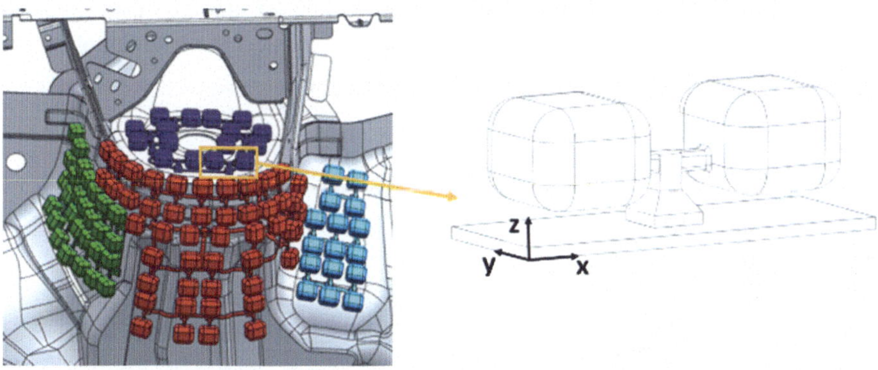

Fig. 4. UC of the LRM solution, consisting of the host structure and the resonant element.

Because of the out-of-plane and in-plane modes of the resonant element, two stop bands open in the same frequency range. A complete stop band for bending waves opens up from 193.1 Hz to 224.1 Hz and a directional stop band opens up from 193 Hz to 224 Hz for those in-plane waves that have a displacement component parallel to y direction.

In previous works [9–13], LRM systems were designed to target only the bending waves, since these are most relevant in view of acoustic radiation. Since the geometry of the shock tower is complex, in-plane and out of plane will couple with each other due to junctions of flat and curved plates with different angles [28], and therefore a LRM is designed, which will attenuate both in-plane as out of plane motion. The importance of blocking both wave types will be demonstrated in Sect. 6.3.

4 Numerical Model of the Vehicle Body and Interior Compartment with NVH solutions

An FE model of the SUV is used to numerically benchmark the designed LRM solution against the TVA and to validate the test data.

4.1 Numerical Model of the Vehicle

The FE model consists of the assembly of the right half of the vehicle's body in white (BIW) and the air volume of the vehicle's interior compartment (see Fig. 5).

The BIW consists of 37 components modeled with a total of 127564 linear shell CQUAD4 elements. The welding connection between these components is modeled with rigid RBE3 connections [25]. The material properties of steel [21] are used for all the components. The components have different thicknesses that vary from 0.8 mm to 2.4 mm: the different colors in the BIW, depicted in Fig. 5, represent these different component thicknesses.

The air volume of the interior compartment is modeled with 82968 linear solid CQUAD8 acoustic elements. The material properties of the air with a density $\rho = 1.205\,\mathrm{kg/m^3}$ and speed of sound $c = 344\mathrm{m/s}$ are used. 1% of damping in air is considered by including a complex speed of sound.

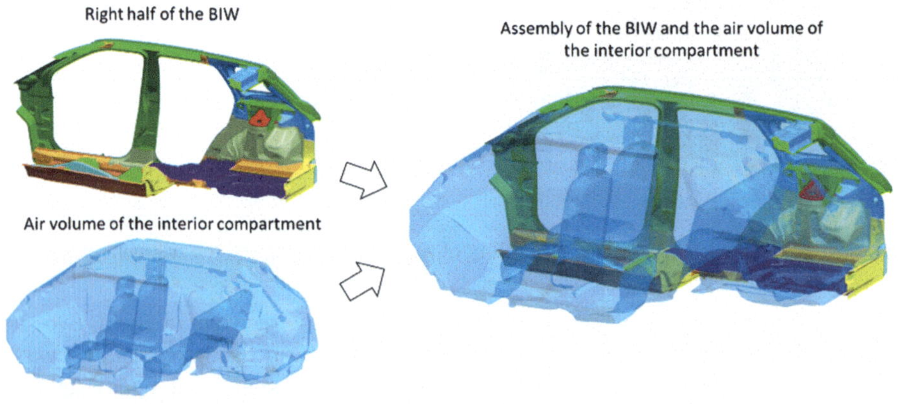

Fig. 5. FE model of the vehicle body and air volume of the vehicle's interior compartment.

The structural and acoustic domains are weakly coupled using the fluid-structure interface ACMODL Nastran card [25]. Nodes of the BIW along the longitudinal axis of the vehicle are clamped.

The SPL is evaluated at the driver right ear (D), at the front passenger left ear (FP) and at the rear right passenger left ear (RP), as sketched in Fig. 6. In view of having a better comparison between the numerical and experimental results, the SPL evaluation locations are chosen close to the longitudinal axis of the vehicle, in order to minimizing the influence of other noise sources, such as windows, which are present in the test campaign but not modeled in the FE model. A Root Mean Square (RMS) value of the SPL is evaluated on an acoustic volume around the ear locations using 4 nodes per ear.

A force vector is applied on each of the four connection points of the shock absorber to the shock tower. The magnitude of the force vector in x, y and z direction is assumed proportional to the acceleration spectra, acquired by a tri-axial accelerometer (PCB Model 356A15) placed on top of the shock tower (see Fig. 6). This allows

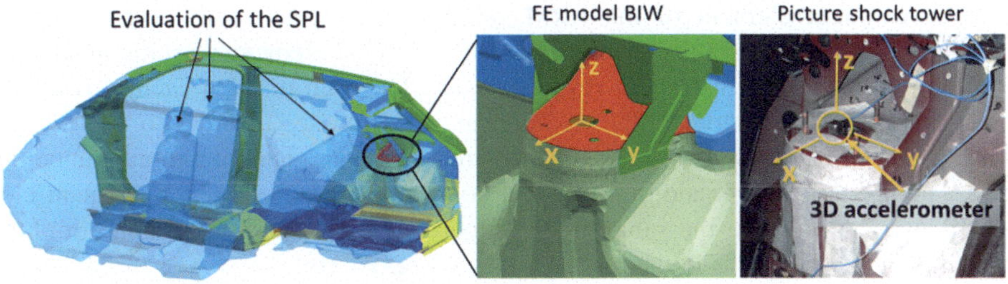

Fig. 6. Evaluation location of the SPL and picture of the tri-axial accelerometer on the right shock tower to acquire the acceleration spectra in x, y, and z directions to be used in the FE model.

to simulate the real dynamic behavior of all the parts connected to the shock tower which are not modeled in the FE model, resulting in excitation conditions closer to the tests conditions.

A modal frequency response solution SOL111 [25] is used to solve the vibro-acoustic problem, in the frequency range from 100 Hz to 300 Hz with a resolution of 1 Hz. The modal base used for the structural domain and acoustic domain includes, for both, all modes from 0 Hz until 1800 Hz.

4.2 Modelling of the TVA and of the Resonant Elements of the LRM Solution

In order to reduce the computational cost of the simulations, both the geometry of the TVA and of the resonant elements are simplified, while the dynamic properties are retained.

The TVA is modeled with two 1D spring CBUSH elements [25] in parallel connected to the two connections points of the shock tower to the shock absorber and one 0D concentrated mass CONM2 element [25] of 1.46 kg and (see Fig. 7 – left). The mode shape of the TVA at 190 Hz is along x direction, on the xy plane. This direction is experimentally measured using a roving hammering approach and 5 lightweight 1D accelerometers (PCB Model 352A24) on the TVA installed on the shock tower.

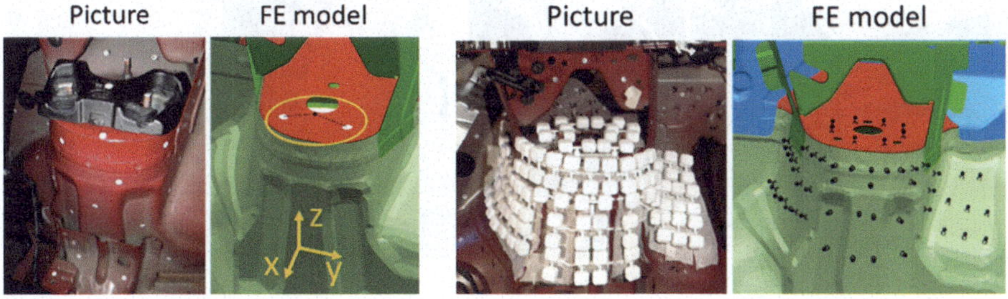

Fig. 7. (Left) Picture and FE model of the TVA. (Right) Picture and FE model of the LRM solution.

Each resonant element of the LRM solution is modeled with one 1D spring CBUSH element and two 0D concentrated mass CONM2 elements: a static mass connected on the host structure and a resonant mass. The total mass of each of resonant element is distributed between the two concentrated masses in order to achieve the same stop band behavior as predicted for the real resonant element and guarantee the correct total (static) mass addition. A local coordinate system for each resonant element is used to preserve the same mode shape orientation direction for the in-plane and out-of-plane modes as in the real layout (see Fig. 7 – right).

5 Measurement Setup

Two different tests are performed: in-lab to have a controlled excitation and on-road to assess actual performance in operation conditions. The steady state harmonic response conditions are considered for both tests.

5.1 In-lab Measurement Setup

The vehicle is excited using two electrodynamic shakers (the Modal Shop 2100E11), which are installed under the vehicle and connected to the bottom of the rear suspensions through a stinger and an impedance head (PCB Model 288D01), as depicted in Fig. 8 – left. A white noise signal is applied in a frequency range from 20 Hz to 512 Hz with 0.5 Hz of resolution.

One tri-axial accelerometer is placed on top of each shock tower, as introduced in Sect. 4.1 and shown in Fig. 6, to measure the acceleration spectra in x, y and z directions upon the given force excitation.

The SPL is measured in the interior compartment using three microphones (PCB Model 378B02). The microphones are located at D, FP and RP locations, as explained in Sect. 4.1. The microphones are placed between the head restraints and the seat back, as shown in Fig. 8 – right.

The air temperature in the interior compartment is monitored during the tests and it is set at 20 °C. The pressure level of the front and rear tires is set at 2.1 atm.

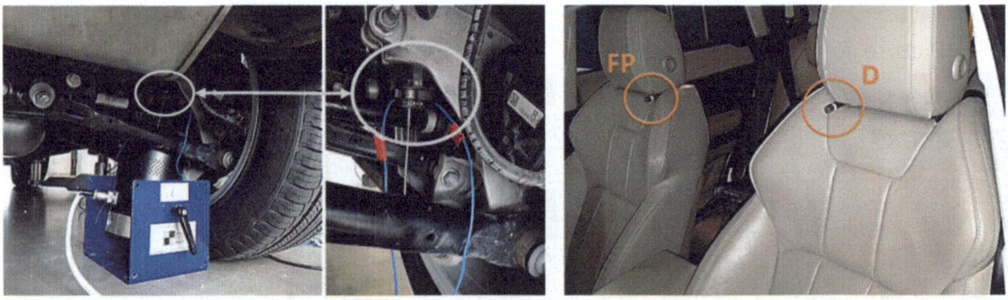

Fig. 8. (Left) Picture of the excitation setup connected under the rear right suspension of the vehicle. (Right) Picture of the microphone positions for the driver and front passenger in the interior compartment.

Frequency Response Functions (FRFs) pressure/force are acquired. Fifty averages are taken to achieve a coherence above 0.9 in the frequency range of interest.

5.2 On-road Measurement Setup

The on-road tests are made on a public road with smooth dry asphalt. Because of the presence of low traffic, no pass-by-noise influence occurred during the acquisitions and a constant driving speed of 50 kph, within a tolerance of ±5%, could be maintained.

For the on-road tests, the same tire pressure level, interior compartment temperature, microphone and accelerometer locations are used as for the in-lab tests. The excitation level, at which the vehicle is subjected during the driving tests, is measured by two tri-axial accelerometers, of same model used on top of the shock towers, which are installed at the location of the impedance heads used in the in-lab tests.

For the on-road tests, the Power Spectral Densities (PSDs) of the pressure and the acceleration are acquired. Each measurement is repeated five times and 150 averages were required to obtain a converged spectrum profile. In order to obtain an average PSD spectrum, for each sensor a RMS of the PSD across the five measurements is calculated.

6 Numerical and Experimental Results

This section reports and discusses the numerical and experimental results for the following three configurations:

- Bare: no NVH treatment installed on the rear shock towers
- TVA: the TVAs are installed on the rear shock towers
- LRM: the TVAs are replaced with the patches

All the SPL results are shown in dBA and the dB reference value is 20e-6 Pa. The graphs are shown in the frequency range between 100 Hz and 300 Hz to better visualize the different between the three configurations in the frequency range of interest.

The acceleration spectra measured on top of the right shock tower in bare configurations are taken as reference and used to model the excitation in the FE model, as explained in Sect. 4.1. The simulations are run for all the configurations, using the load excitation obtained from the acceleration spectra of the in-lab tests and of the on-road tests.

6.1 In-lab Results

In order to compare the numerical and experimental results, the FRF SPL test results are shown only for the source of excitation acting on the rear right shock tower. The test results for the excitation acting on the rear left shock tower can be found in [21].

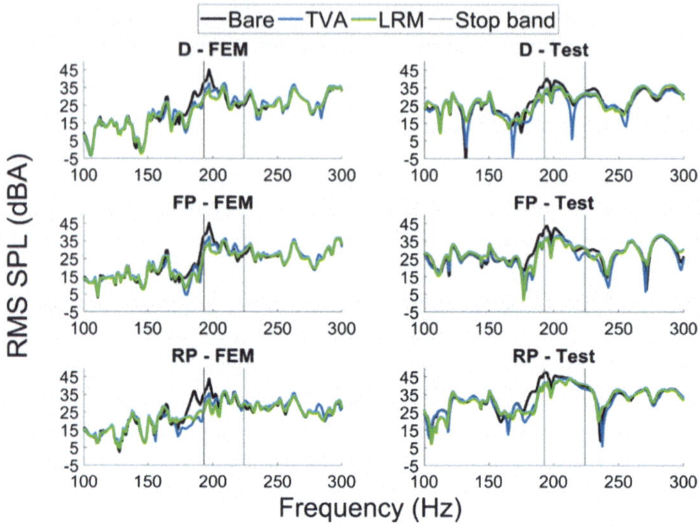

Fig. 9. Comparison between the numerical prediction and experimental results for the in-lab tests. (Left column) Numerical results. (Right column) Experimental results.

A peak at 195 Hz can be observed from the bare results and it is well predicted in the numerical model. The numerical and experimental results shown in Fig. 9 shows a similar trend, which can be summarized as follows:

- The SPL of the TVA configuration is lower than the bare configuration in the frequency range from 185 Hz to 210 Hz.
- Outside this frequency range, the SPL of the TVA and bare configurations are similar.
- The SPL of the LRM configuration performs similar as the TVA configuration for the frequency range from 185 Hz to 210 Hz.
- The LRM configuration shows a strong reduction of the SPL around the lower bound of the predicted stop band. The energy injected into the vehicle body is strongly attenuated throughout all the stop band frequency range, but achieving a maximum attenuation at the lower bound, since the wave attenuation close to the lower stop band bound is maximum [23].

6.2 On-road Results

A similar level of excitation is used for all the tested configurations, as depicted in Fig. 10 by the RMS PSD acceleration measured at the bottom of the rear suspensions. The consistency of the results between different measurements of the same configuration is confirmed by a maximum standard deviation value of the RMS PSD SPL varying within ± 1.01 dBA in the frequency range of interest.

A pronounced peak at 198 Hz in the acceleration levels is measured in the on-road tests. This peak is also measured and well predicted by the numerical model in the RMS PSD SPL in the interior compartment.

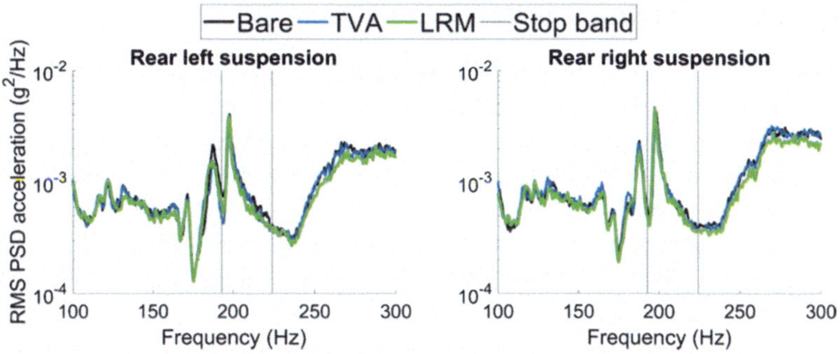

Fig. 10. RMS PSD acceleration measured at the bottom of the rear left and right suspensions.

The improvement in the SPL for the TVA and LRM configuration is in a similar frequency range as in the in-lab tests and it is confirmed by the numerical predictions (see Fig. 11). In the on-road tests, the LRM configuration shows to perform better than the TVA configuration at 198 Hz.

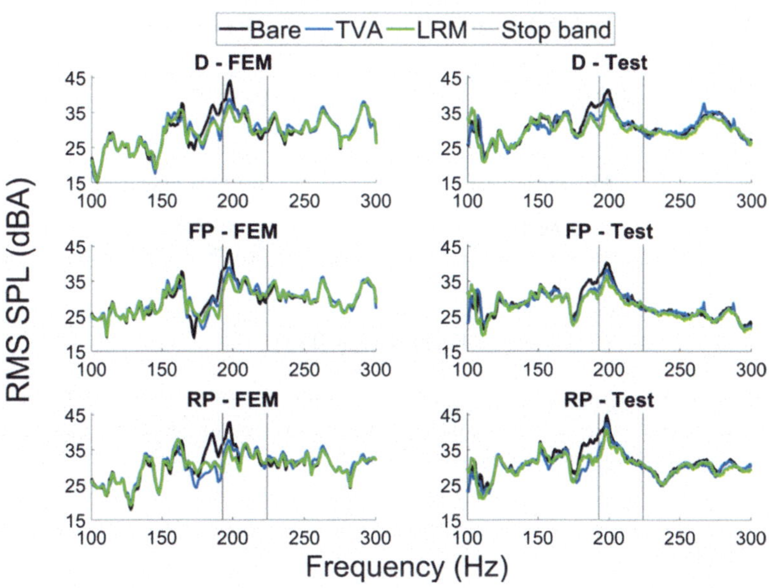

Fig. 11. Comparison between the numerical prediction and experimental results for the on-road tests. (Left column) Numerical results. (Right column) Experimental results.

6.3 Importance of Multi-mode Resonant Elements

This section shows the difference in SPL predicted by the proposed LRM system that leads to a stop band for both out-of-plane and in-plane waves (LRM OOP+IP) and a LRM system leading to a stop band only for out-of-plane waves (LRM OOP). Numerical predictions are depicted in Fig. 12, using load data obtained from the acceleration spectra of the in-lab tests and of the on-road tests.

The numerical results in Fig. 12 show that the LRM OOP+IP has up to 2 dB more attenuation in SPL than the LRM OOP in the frequency range from 185 Hz to 210 Hz. Outside this frequency range, the two LRM systems have similar performance. These results confirm the importance of designing a LRM system that achieves stop band behavior for more than one wave type for this kind of application with a high geometrical complexity.

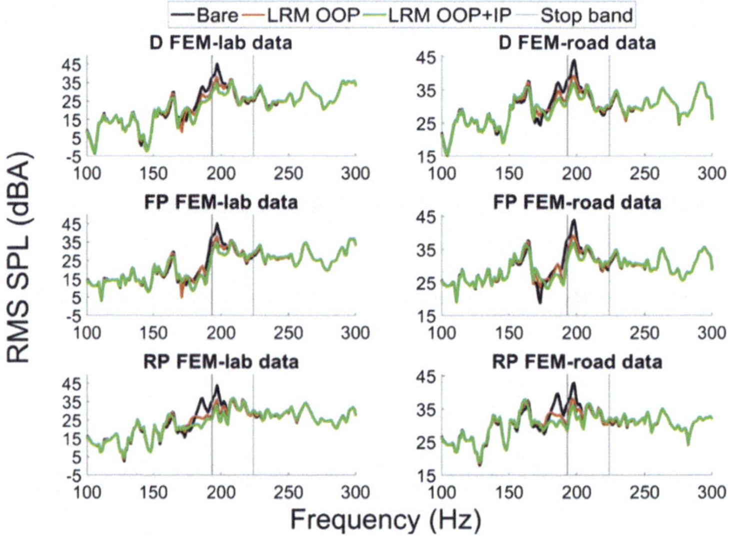

Fig. 12. Numerical comparison of the RMS SPL results for LRM OOP and LRM OOP+IP configurations. (Left column) Excitation spectra from the in-lab tests. (Right column) Excitation spectra from the on-road tests.

7 Conclusion

In this paper, a current NVH solution in a commercial vehicle is replaced by a lighter LRM concept with the goal of reducing the noise in the interior compartment of a vehicle, caused by tire/road interaction while driving. The LRM concept consists of patches composed of resonant elements that are applied around the rear shock towers of the vehicle, where a TVA is already present, in order to reduce the vibrational energy being injected into the vehicle body through the suspension assembly system around 190 Hz.

The LRM configuration is compared with the bare and TVA configurations. Using a simplified FE model of the vehicle, the experimental findings made in-lab and on-road are numerically verified. The LRM configuration achieves similar SPL performance in the interior compartment as the TVA configuration in the in-lab tests, whereas it performs better in the on-road tests. Numerical results confirm the importance of designing a LRM solution that achieves stop band behavior for more than one wave type in case of a complex geometry of the host structure, in order to reduce effectively the transmission of energy along the treated transfer paths and achieve better SPL attenuation in the target frequency range.

In this hard-to-address low frequency vibro-acoustic problem, the LRM solution demonstrates to be a valid alternative to the current NVH solution, by having comparable SPL performance in the interior compartment and by reducing the added mass by 48%.

Acknowledgments. The Research Fund KU Leuven is gratefully acknowledged for its support. This research was partially supported by Flanders Make, the Strategic Research Centre for the manufacturing industry. The research of E. Deckers is funded by a grant from the Research Foundation – Flanders (FWO).

References

1. Mallick, P. K. Materials, design and manufacturing for lightweight vehicles. *Elsevier* (2010).
2. Fritschi, L., Brown, A., Kim, R., Schwela, D., Kephalopolous, S. Burden of disease from environmental noise: quantification of healthy life years lost in Europe. *World Health Organization, Regional Office for Europe, Copenhagen* (2011).
3. Fahy, F. J., Gardonio, P. Sound and structural vibration: radiation, transmission and response. *Elsevier* (2007).
4. L. Brillouin, Wave propagation in periodic structures: electric filters and crystal lattices, *Courier Corporation* (2003).
5. C. Goffaux, J. Sánchez-Dehesa, A. L. Yeyati, P. Lambin, A. Khelif, J. Vasseur, B. Djafari-Rouhani, Evidence of fano-like interference phenomena in locally resonant materials, *Physical review letters* 88 (22) (2002).
6. Z. Liu, X. Zhang, Y. Mao, Y. Zhu, Z. Yang, C. T. Chan, P. Sheng, Locally resonant sonic materials, *Science* 289 (5485), pp 1734–1736 (2000).
7. F. Lemoult, N. Kaina, M. Fink, G. Lerosey, Wave propagation control at the deep subwavelength scale in metamaterials, *Nature Physics* 9 (1) 55 (2013).
8. X. Zhou, X. Liu, G. Hu, Elastic metamaterials with local resonances: an overview. *Theoretical and Applied Mechanics Letters* 2 (4) (2012).
9. Claeys, C. C., Deckers, E., Pluymers, B., & Desmet, W. A lightweight vibro-acoustic metamaterial demonstrator: numerical and experimental investigation. *Mechanical Systems and Signal Processing*, 70, 853–880 (2016).
10. De Melo Filho, N. G. R., Van Belle, L., Claeys, C., Deckers, E., Desmet, W. Dynamic mass based sound transmission loss prediction of vibro-acoustic metamaterials double panels applied to the mass-air-mass resonance. *Journal of Sound and Vibration*, 442, 28–44 (2019).

11. De Melo Filho, N. G. R., Sangiuliano, L., Claeys, C., Deckers, E., Desmet, W. Vibro-acoustic metamaterials for increased STL in acoustic resonance driven environments. *Proceedings of NOVEM*, 257 (1), pp. 537–549, Ibiza, Spain (2018).
12. Claeys, C., Melo, N. F., Van Belle, L., Deckers, E., & Desmet, W. Design and validation of metamaterials for multiple structural stop bands in waveguides. *Extreme Mechanics Letters, 12*, 7—22 (2017).
13. Sangiuliano, L., Claeys, C., Deckers, E., Pluymers, B., Desmet, W. Force isolation by locally resonant metamaterials to reduce NVH. *SAE Technical Paper Series,* 2018-01-1544 (2018).
14. Douville, H., Masson, P., Berry, A. On-resonance transmissibility methodology for quantifying the structure-borne road noise of an automotive suspension assembly. *Applied Acoustic 67,* 358–382 (2006).
15. Kindt, P., Berckmans, D., De Coninck, F., Sas P., Desmet, W. Experimental analysis of the structure-borne tyre/road noise due to road discontinuities. *Mechanical Systems and Signal Processing 23*, 2557–2574 (2009).
16. Tatlow, J., Ballatore, M. Road noise input identification for vehicle interior noise by multi-reference Transfer Path Analysis. *Procedia engineering 199*, 3296–3301 (2017).
17. T. Sakata, H. Morimura, H. Ide, Effects of Tire Cavity Resonance on Vehicle Road Noise. *Tire Science and Technology*, 18 (2), pp. 68–79 (1990).
18. Hartleip, L. and Roggenkamp, T. Case Study – Experimental Determination of Airborne and Structure-borne Road Noise Spectral Content on Passenger Vehicles. *SAE Technical Paper* 2005-01-2522 (2005).
19. Kido, I., Nakamura, A., Hayashi, T., and Asai, M. Suspension Vibration Analysis for Road Noise Using Finite Element Model. *SAE Technical Paper* 1999-01-1788 (1999).
20. Kim, G.J., Holland, K.R. and Lalor, N. Identification of the airborne component of tyre-induced vehicle interior noise. *Applied Acoustics*, 51 (2), pp. 141–156 (1997).
21. Sangiuliano, L., Claeys, C., Deckers, E., Pluymers, B., Desmet, W. Reducing Vehicle Interior NVH by Means of Locally Resonant Metamaterial Patches on a Rear Shock Tower. *SAE Technical Paper Series,* 2019-01-1502 (2019).
22. Wang, G., Wen, X., Wen, J., Shao, L., & Liu, Y. Two-dimensional locally resonant phononic crystals with binary structures. *Physical review letters, 93*(15), 154302 (2004).
23. Claeys, C., Vergote, K., Sas, P., & Desmet, W. On the potential of tuned resonators to obtain low-frequency vibrational stop bands in periodic panels. *Journal of Sound and Vibration*, 332(6), 1418–1436 (2013).
24. Brillouin, L. Wave propagation in periodic structures (Vol. 2nd). *McGraw-Hill Book Company* (1946).
25. Siemens, NX Nastran 12. Quick Reference Guide, *Siemens PLM Software Inc.* (2017).
26. B. R. Mace, E. Manconi. Modelling wave propagation in two-dimensional structures using finite element analysis, *Journal of Sound and Vibration*, 318 (4), pp. 884–902 (2008).
27. Langley, R. S. A note on the force boundary conditions for two-dimensional periodic structures with corner freedoms. *Journal of Sound and Vibration*, 162(2), 377–381 (1993).
28. Cremer, L., Heckl, M., Petersson, B.A.T. Structure-borne sound. *Springer*, 341–359 (2005).

Design of Sound Package for Pass-By Noise Reduction: Process and Application

Claudio Bertolini(✉), Jan Horak, and Thibault Lafont

Autoneum Management AG, Schlosstalstrasse 43, 8406 Winterthur, Switzerland
claudio.bertolini@autoneum.com

Abstract. The latest Pass-By (PB) noise regulation ECE R51.03 requires OEMs to achieve ever more challenging limits. In particular, the target of 68 dB(A) in 2024 set by the ECE R51.03 Phase III appears as a critical challenge, considering that such limit has to be achieved with the updated ISO 362-1:2015 measurement method. With this new measurement method, the contributions to PB noise given by tires and powertrain are more balanced compared to the past, when powertrain was the dominant source. This fact makes the achievement of the above mentioned demanding levels inherently more difficult.

As a matter of fact, for the achievement of PB noise limits OEMs are now compelled to define quantitative strategies for the reduction of powertrain and tire noise separately. This must start from an accurate assessment of the relative share that these two sources have in the PB noise level, then pass to the definition of corresponding reduction targets and eventually design concrete countermeasures to achieve them. In particular, in this last step, vehicles are to be fitted with some optimized exterior acoustic package, which requires a thorough insight into the noise radiation from both powertrain and tires.

This article is subdivided into two parts. In the first part, the mentioned process is elaborated in some detail, with special emphasis on important practical aspects, e.g. how to define sensible noise reduction targets for tire and powertrain noise. In the second part of the article, an application example is shown, in which the process is put in place and the package of a B-segment European vehicle is optimized by means of different indoor and outdoor testing procedures, allowing the achievement of the 68 dB(A) PB noise level.

Keywords: Pass-By noise · ISO 362 · Tire noise · Powertrain noise

W. Siebenpfeiffer (Hrsg.): *Automotive Acoustics Conference 2019*, Proceedings, S. 137–169, 2020. https://doi.org/10.1007/978-3-658-27669-0_11

List of Abbreviations

PB Pass-By
PMR Power to Mass Ratio
NVH Noise vibration and harshness
SCA Source Contribution Analysis
FRN Free Rolling Noise
TI Torque Influence
SPL Sound Pressure level
ATF Acoustic Transfer Functions
SMT Substitution Monopole Technique

1 Introduction

The ISO-362 measurement standard has represented for many years the back-bone of all legislations on exterior noise of transportation vehicles, not only in Europe (where it was implemented in regulation ECE R51.02). This standard was first introduced in 1975, with the purpose of defining a procedure for the assessment of exterior noise that could be representative of the "noisiest urban driving conditions". The measurement procedure prescribed by the ISO-362 consisted in driving the vehicle on a 20 m-long straight test-track in full acceleration in 2nd and 3rd gear starting from a speed of 50 km/h and simultaneously measuring the noise generated at two microphones placed at 7.5 m from its longitudinal axis and at 1.2 m from the ground (the so-called Pass-By noise), mid-way along the test-track (see Fig. 1).

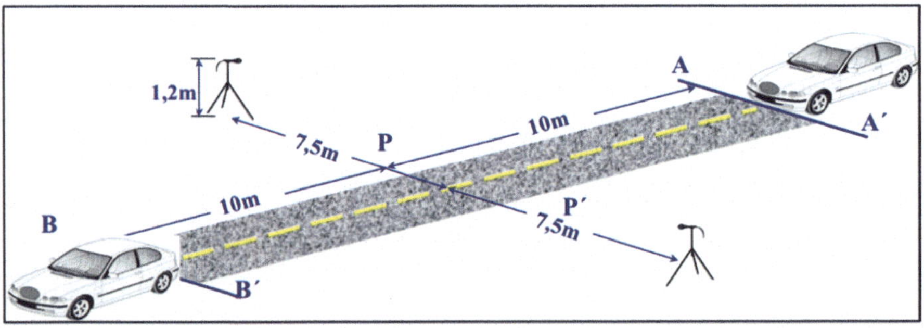

Fig. 1. Topological layout of a Pass-By test according to ISO 362

While the "geometry" of measurement procedure described in the ISO 362 was never put in discussion, already in the 90s [1] it was recognized that the driving conditions prescribed by it were not really representative of the "noisiest urban driving conditions", since wide-open-throttle acceleration is only rarely observed in urban traffic. In terms of noise sources, these driving conditions were too much privileging the powertrain noise contribution, at the expense of tire noise which is, in several cases, far from being negligible in urban driving conditions. This predominance of powertrain noise in ISO 362 tests was further favoured by the fact that many regulatory implementations of this standard allowed the use of tyres with minimum tread depth. On top of all these issues related to the driving conditions, the technical development of vehicle engine technologies and transmission technologies had made some of the original technical assumptions behind the first ISO 362 to no longer be valid.

For these reasons, a revision process was triggered, which led to a revised version of the standard (named ISO 362-1), a first draft of which was published in 2007 and that found its final form in the ISO 362-1:2015 [2]. In this norm, the prescribed driving conditions are very different from those of the initial ISO 362. More specifically, according to the new measurement method the final PB noise level is a combination of the noise emitted by the vehicle in a wide-open-throttle (wot) acceleration and at a constant speed. For both driving conditions, suitable gears must be chosen in such a way that, by taking the above mentioned combination, what is obtained in the end is an estimate of the noise emitted by the vehicle during a partial-load acceleration at a specific target value a_{urban} of the acceleration itself and at 50 kph since, from statistical data, these appear to be the "noisiest typical urban driving conditions" [2–4]. With this new way of assessing Pass-By noise, the contribution of tires to the final PB noise value turns out to be higher than with the previous measurement procedure, since it is known that tire noise plays an important role at constant speed and for partial load accelerations. As a consequence, with the new driving conditions *the contributions from the tires and from the powertrain to the PB noise level will be -generally, at least-more balanced*. Furthermore, *such contributions will also be more vehicle dependent*, as a result of the way noise levels measured at constant speed and in wot acceleration are "mixed" in the ISO 362-1, which is something (strongly) vehicle dependent.

In Europe, all these changes in the exterior noise measurement procedure were incorporated in regulation ECE 51.03 -issued in 2014- in combination with a road-map for a progressive reduction of the allowable PB noise level. Such level, currently at 72 dB(A), will be reduced to 70 dB(A) in 2020 and to 68 dB(A) in 2024 (see Fig. 2).

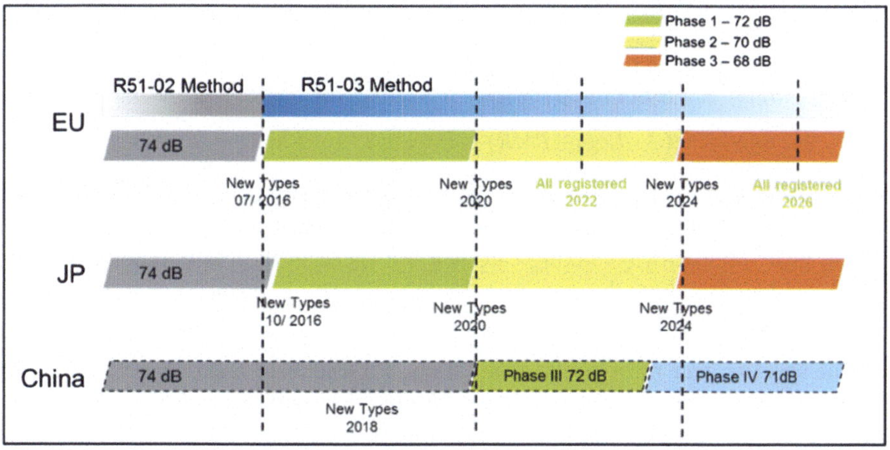

Fig. 2. Road map for implementation of PB noise limits in EU, Japan and China

A reduction of 4 dB(A) in PB noise in a time span of a few years represents an important challenge for NVH engineers dealing with exterior noise, in particular when such challenge comes along with a change in the measurement procedure that makes the shares of PB noise due to powertrain and due to tires more unpredictable and less easy to master. This challenge is first of all on the shoulders of the car makers but -of course- it is in a way or the other cascaded down by them to the other stakeholders involved, namely the tire manufacturers and the suppliers of passive solutions for noise reduction.

This article intends to describe some of the work that was done in this field by a sound package supplier during the last 3 years, with the purpose of better understanding how to handle the higher complexity of the new ISO 362-1, how to define sensible targets for PB powertrain and tire noise reduction and how to implement solutions that allow achieving such targets.

2 A Quick Review of the ISO 362-1:2015

The purpose of this section consists in reviewing in detail the little "algebra" at the basis of the evaluation of the PB noise level as it is prescribed by the ISO 362-1. In the intent of this standard, all this algebra is aimed at getting the noise level made by a vehicle at the speed of 50 kph and subject to a specific *partial-load* acceleration a_{urban} that depends on the vehicle Power to Mass Ratio (PMR) and that is given by the following formula:

$$a_{URBAN} = 0.63 \cdot Log_{10}(PMR) - 0.09\,\mathrm{m/s^2} \tag{1}$$

These very peculiar driving conditions are selected in the ISO 362-1 since statistical data seem to indicate that they represent the "noisiest urban driving conditions" [2–4]. Since partial load tests are not easily repeatable, in the ISO 362-1 it is proposed to mimic them by a combination of wot acceleration tests and constant speed tests.

In order to define the gears at which all these tests must be carried out, ISO 362-1 further introduces the acceleration a_{ref}, which is different from a_{urban} and depends on the vehicle PMR too:

$$\begin{aligned} a_{REF} &= 1.59 \cdot Log_{10}(PMR) - 1.41\,\mathrm{m/s^2} \text{ for PMR } > 25\,\mathrm{W/kg} \\ a_{REF} &= 0.63 \cdot Log_{10}(PMR) - 0.09\,\mathrm{m/s^2} \text{ for PMR } < 25\,\mathrm{W/kg} \end{aligned} \tag{2}$$

Given this definition of a_{ref}, in order to choose the gears at which ISO 362-1 tests should be carried out, it is necessary to run "preliminary wot acceleration tests" in different gear ratios (always taking care that the speed at the PB microphones line is 50 km/h) and analyse how the average vehicle acceleration during such tests compares to a_{ref}. More specifically:

1. If one specific gear ratio gives acceleration in a tolerance band of ±5% of the reference acceleration, a_{ref}, not exceeding 2.0 m/s^2, test with that gear ratio;
2. If no gear ratio gives the required acceleration, then choose a gear ratio i, with an acceleration higher and a gear ratio (i + 1), with an acceleration lower than the

reference acceleration a_{ref}. If the acceleration value in gear ratio i does not exceed 2.0 m/s^2, use both gear ratios for the test. A gear ratio weighting factor k_G is then defined in the following way:

$$k_G = \frac{a_{REF} - a_{wot(i+1)}}{a_{wot(i)} - a_{wot(i+1)}} \tag{3}$$

3. If the acceleration value of gear ratio i or (i + 1) exceeds 2.0 m/s^2, the first gear ratio shall be used that gives an acceleration below 2.0 m/s^2 unless gear ratio (i + 1) provides acceleration less than a_{urban}.
4. In the case where gear ratio (i + 1) provides acceleration less than a_{urban}, two gears, i and (i + 1), shall be used, including the gear i with acceleration exceeding 2.0 m/s^2. The gear ratio weighting factor k_G is then calculated using formula (3).

According to these prescriptions, car must be tested either in two different gear ratios (cases 2 and 4) or in one single gear ratio (cases 1 and 3). We will assume here -for a moment- to consider the case in which 2 gear ratios must be tested. In this case four tests are needed in total: two in cruise conditions (at 50 kph) and two in wot acceleration conditions for gear ratios i and i + 1[1]. From these tests, two noise levels L_{wot} and L_{cru} are evaluated for each PB noise microphone in the following way:

$$L_{wot} = k_G \cdot L_{wot(i)} + (1 - k_G) \cdot L_{wot(i+1)} \tag{4}$$

$$L_{cru} = k_G \cdot L_{cru(i)} + (1 - k_G) \cdot L_{cru(i+1)} \tag{5}$$

where the "noise levels" in formulas (4) and (5) must be intended, as usual, as the maximum recorded by the PB microphone during the test. It is important to understand the meaning of formulas (4) and (5). If one assumes that the noise level at a PB microphone in wot acceleration is a *linear function of the average acceleration during the test*, formula (4) gives an estimate of such noise level for acceleration = a_{REF}. Formula (5) is obtained simply as a consequence, since the way of linearly combining constant speed tests (for which it does not make sense to talk about acceleration) should be the same as that used for wot acceleration tests.

Once L_{wot} and L_{cru} are given, for each PB microphone a level L_{urban} is evaluated in the following way:

$$L_{urban} = k_P \cdot L_{cru} + (1 - k_P) \cdot L_{wot} \tag{6}$$

where k_P is a "partial load factor" given by:

$$k_P = \frac{a_{REF} - a_{URBAN}}{a_{REF}} \tag{7}$$

---

[1] All levels reported in formulas (4) and following are meant to represent average values over 4 repetitions of each test, according to the prescriptions of ISO 362-1.

Formula (6) is again a linear interpolation, that aims at estimating the noise level at the PB microphones for acceleration = a_{urban} under the assumption that such noise level is a linear function of the average acceleration during the test, at least over the acceleration range $[0 - a_{ref}]$. In the end the "PB noise level" is the maximum of the two L_{urban} levels evaluated at the two PB noise microphones, rounded to the nearest integer value.

Combining formulas (4)–(6) it is easy to see that:

$$\begin{aligned} L_{urban} = {} & k_P \cdot k_G \cdot L_{cru(i)} + k_P \cdot (1 - k_G) \cdot L_{cru(i+1)} \\ & + (1 - k_P) \cdot k_G \cdot L_{wot(i)} + \left(1 - k_p\right) \cdot (1 - k_G) \cdot L_{wot(i+1)} \end{aligned} \tag{8}$$

Formula (8) shows in more explicit form that the level L_{urban} is a linear combination of the noise levels evaluated in the 4 driving conditions to be considered according to the prescriptions of ISO 362-1. And given that the sum of the coefficients is obviously equal to one, this linear combination is actually a "weighted average", where each level is weighted according to a specific coefficient that can be interpreted as a "contribution" to L_{urban}.

Just as an example, we can consider here the case of a B-segment vehicle that was the object of an acoustic study carried out in collaboration with a car maker during year 2018 [5]. For such vehicle, PMR = 48.5 kW/t, $a_{urban} = 0.97\ m/s^2$, $a_{REF} = 1.27\ m/s^2$ and $k_p = 0.235$. From preliminary acceleration tests it turned out that the car had to be tested in 3rd and 4th gear and the value of k_G from formula (3) turned out to be 0.73. For such a case, formula (8) reads:

$$L_{urban} = 0.17 \cdot L_{cru(3)} + 0.06 \cdot L_{cru(4)} + 0.56 \cdot L_{wot(3)} + 0.21 \cdot L_{wot(4)} \tag{9}$$

which shows that the main "contributor" to L_{urban} is by far $L_{wot(3)}$ with its 56%, while $L_{cru(4)}$ is the smallest contributor, with only 6%. Formula (9) shows also that levels measured in wot accelerations contribute for 77% to L_{urban}, while levels measured in cruise conditions contribute only 23% to L_{urban}.

A similar analysis, though simpler, can be carried out also in case only one gear ratio has to be tested. In such a case, for the evaluation of L_{urban}, formula (6) is still applied, but with k_p given by:

$$k_P = \frac{a_{wot} - a_{URBAN}}{a_{wot}} \tag{10}$$

where a_{wot} is the average acceleration measured during the wot acceleration test. Here, as an example we can consider the case of a European D-segment vehicle with PMR = 101 kW/t and $a_{urban} = 1.17\ m/s^2$. From preliminary acceleration tests, it turned out that the vehicle had to be tested only in 4th gear and k_p calculated from formula (10) turned out to be equal to 0.28. In such a case formula (6) reads:

$$L_{urban} = 0.28 \cdot L_{cru} + 0.72 \cdot L_{wot} \tag{11}$$

which shows that the level measured in wot acceleration contributes 72% to L_{urban}, while the level measured in cruise conditions contributes "only" 28%.

This type of analysis, while being extremely simple, helps understanding the importance of the levels measured in the different driving conditions, in relation to the only final figure that matters for the exterior noise regulation, which is L_{urban}. In turn, this fact can be helpful in deciding which direction should be taken to improve this figure, in case this is necessary. This will be the topic of the next sections.

3 Estimation of Contributions to PB Noise from Tires and Powertrain

"Contribution analysis" is a task that is very well known to NVH engineers dealing with exterior noise and PB tests, who are normally required to quantify how much the different sources (tires, engine, exhaust, intake…) contribute to the PB noise level. Historically, for many years this has been done directly on the test-track, through "masking" tests in which PB noise measurements are executed by eliminating (with suitable hardware) each source of interest. More recently, PB noise Source Contribution Analysis (SCA) methods have been proposed that are based on indoor PB measurements [6–10]. One of these methods will be reviewed and considered later on in this paper. In this section a simpler approach is taken, in which it is assumed that the dominant contributions to PB noise are the tires and the powertrain and that other contributions (typically, exhaust and intake) can be neglected. While this might not be true in general, many results reported in the technical literature [6–10] seem to indicate that this is often the case. Also for the EU B-segment vehicle mentioned in previous sections, tests carried out by the car maker [5] showed that the exhaust contribution was only 5%.

In such a situation, carrying out a contribution analysis is equivalent to being able to isolate either the powertrain or the tires contribution. A way to isolate the tire noise contribution is proposed in the standard ISO 362-3, which is the standard about indoor PB tests [11, 12]. According to the procedure proposed in the ISO 362-3, to estimate the tires contribution, two terms must be summed up: The Free Rolling Noise (FRN) and the Torque Influence (TI) of the tire under load. The FRN estimation can be accomplished starting from a set of coast-down tests at different speeds (covering the speed range of PB noise tests). Starting from these data and by means of a logarithmic regression model, it is possible to build a "Free Rolling Noise model", i.e. a function that, for any given velocity profile $v = v(x)$, gives the corresponding $FRN = FRN(x)$, where x is the position of the vehicle along the PB noise track. For the case of the EU B-segment vehicle mentioned in previous sections, outdoor coast-down test results were provided by the car maker. Figure 3 shows their speed profiles, that well cover the speed range of the PB noise tests for the considered vehicle. Figure 4 shows the corresponding SPL profiles.

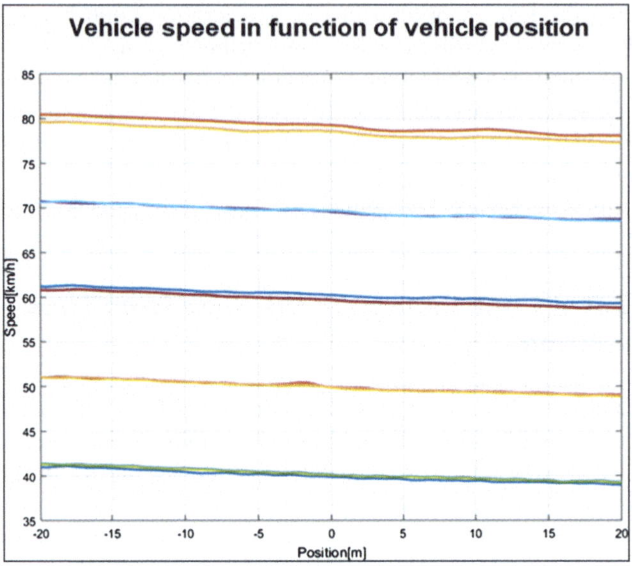

Fig. 3. Coast down tests for estimation of PB Free Rolling Noise – speed profiles

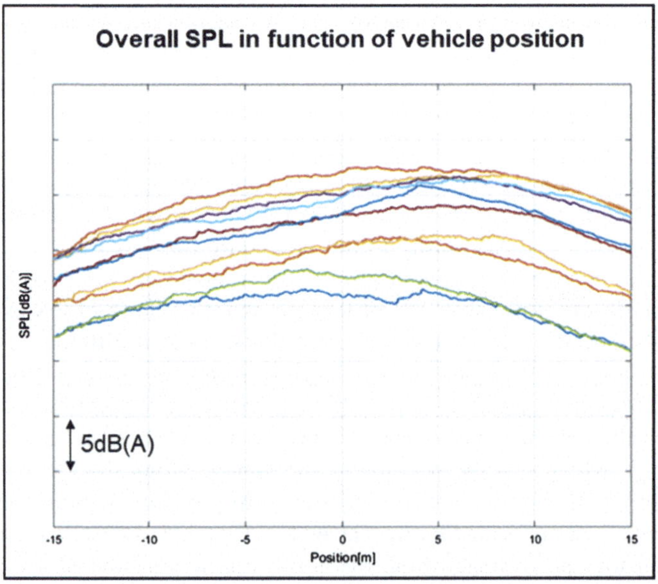

Fig. 4. Coast down tests for estimation of PB Free Rolling Noise – SPL profiles

A logarithmic interpolation of these data is carried out using the formula (taken from ISO 362-3 [12]):

$$L_{FRN}(v,x) = \alpha(x) + \beta(x) \cdot lg\left(\frac{v(x)}{50\,\text{km/h}}\right) \tag{12}$$

where α(x) and β(x) (position dependent, in dB(A)) are estimated with a least-squares procedure, v is the vehicle speed in km/h and x is the vehicle position.

Figure 5 shows a 3d chart of L_{FRN} as a function of vehicle speed and position. From this function it is possible to estimate the FRN for every PB noise driving condition of interest, provided its velocity profile v = v(x) (i.e. the velocity of the vehicle as a function of the position during the test) is known. For example, in Fig. 5 the red line represents the estimation of the FRN for a 3rd gear wot acceleration.

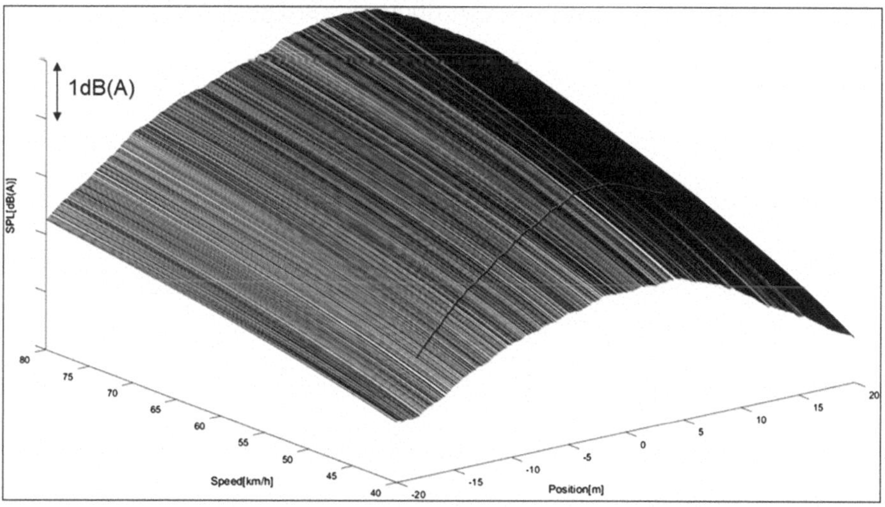

Fig. 5. Free Rolling Noise model as a function of vehicle speed and position – Red line represents a "cut" corresponding to the speed profile of a 3rd gear wot acceleration test

Coming now to the Torque Influence (TI), ISO 362-3 suggests two procedures for its estimation. The first procedure is based on the execution of wot acceleration tests with a "tyre test vehicle", i.e. a vehicle in which the powertrain contribution has been somehow suppressed. By comparing the noise at the PB microphones obtained in these wot acceleration tests with that measured in coast-down conditions, a "Torque Influence model" TI = TI(v, x) is built, again through regression techniques. In case a "tyre test vehicle" is not available, a simplified procedure is proposed, where the TI is assumed to depend on the acceleration in a quadratic way [12]:

$$TI(a,x) = 2 \cdot \zeta \cdot a^2(x) + \zeta \cdot a(x) \tag{13}$$

where a is the acceleration and ζ is a coefficient that depends on the test track and that normally ranges between 0.075 (for quieter test tracks) and 0.15 (for louder test tracks).

For the case of the B-segment vehicle considered here, a "tyre test vehicle" was not available and for this reason, the "simplified method" was used. By taking an average value ford $\zeta = 0.11$ in formula (13), a TI of about 0.56 dB(A) is obtained for 3rd gear wot and a TI of about 0.27 dB(A) is obtained for 4th gear wot.

Figure 6 shows the total PB noise recorded (at the right microphone) during a 3rd gear wot acceleration test in combination with the tire noise contribution evaluated by means of the above described procedure and the powertrain contribution evaluated simply by energetic difference. Similar results are shown in Figs. 7, 8 and 9 for 4th gear wot acceleration, 3rd gear cruise and 4th gear cruise respectively.

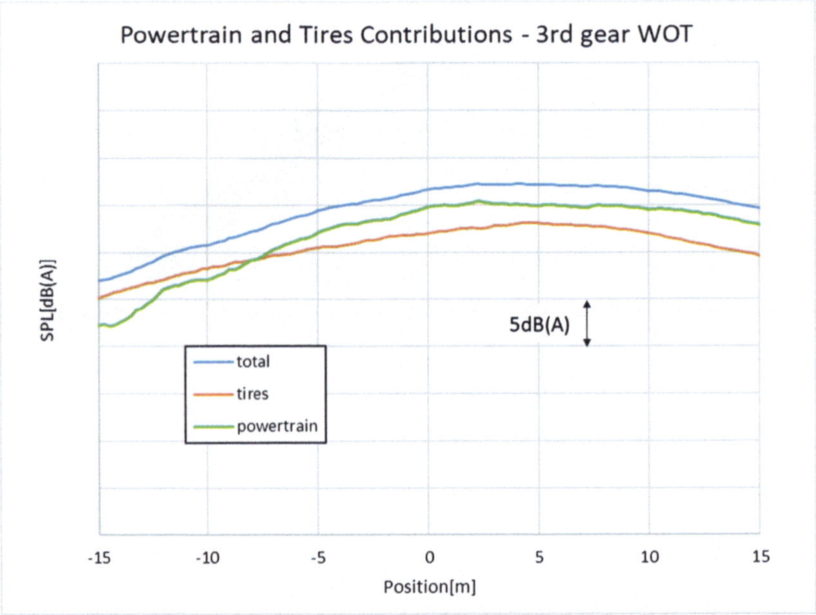

Fig. 6. Powertrain and tires contributions – 3rd gear wot acceleration

All results in these figures, as well all results reported in what follows, refer to the right PB microphone, which was the one recording the highest levels in this specific case.

Of course, contributions reported in all these figures are likely to include some level of approximation. This is due to the assumption initially made according to which PB noise is determined only by powertrain and tires. But it is due also to approximations and uncertainties in the FRN model and in the TI model. This is visible for example in Fig. 9, where between 5 m and 10 m the powertrain contribution suddenly drops to zero, since the calculated tires contribution is basically equal (or even slightly higher) than the total PB noise.

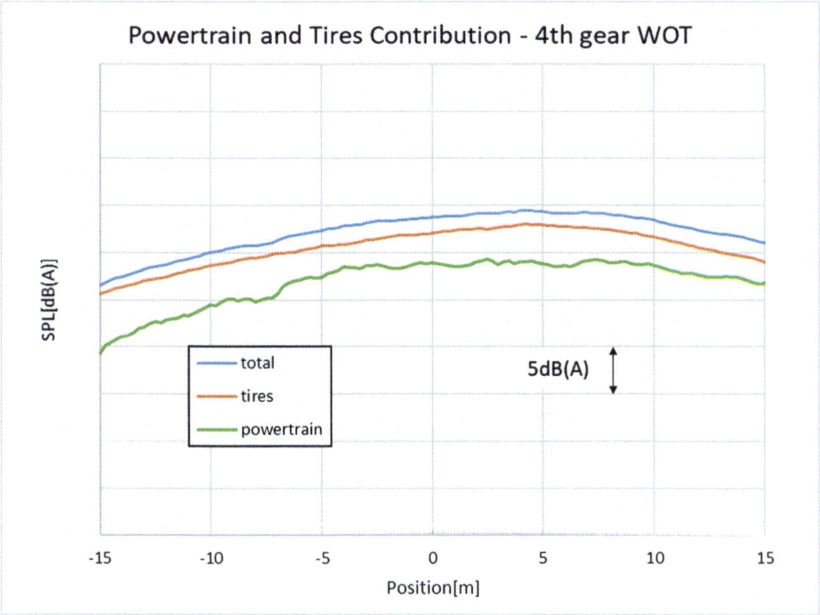

Fig. 7. Powertrain and tires contributions – 4th gear wot acceleration

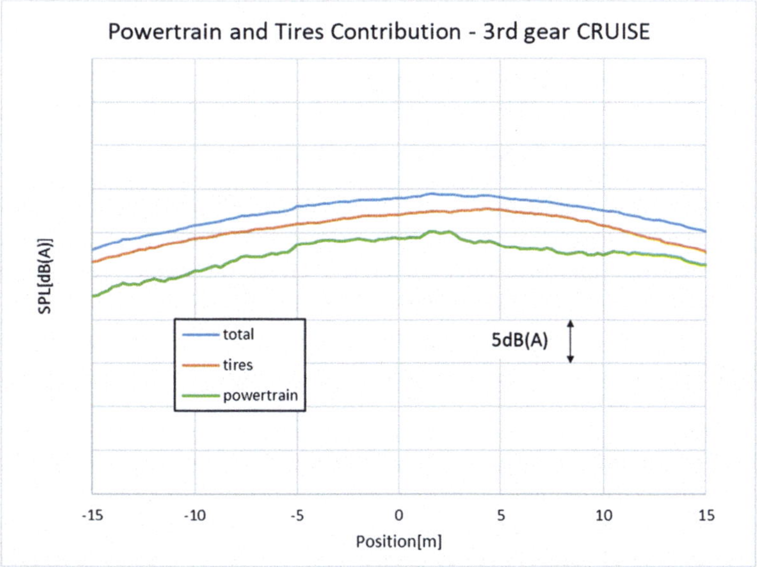

Fig. 8. Powertrain and tires contributions – 3rd gear 50 kph constant speed

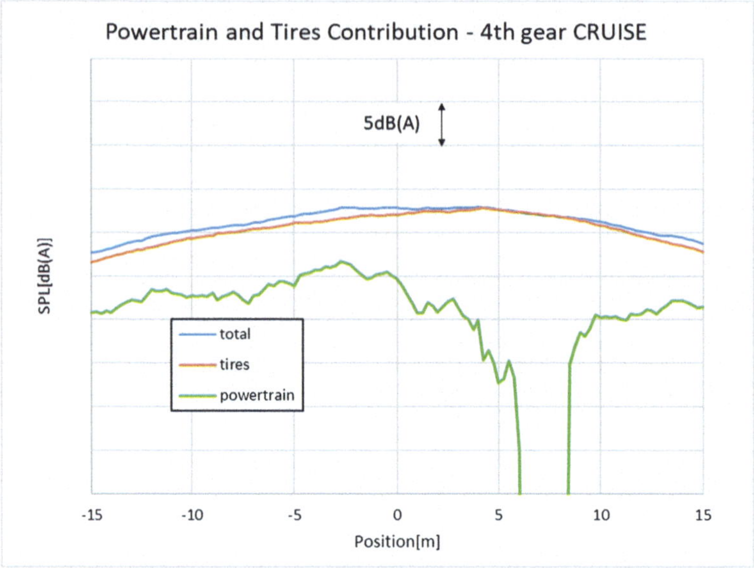

Fig. 9. Powertrain and tires contributions – 4th gear 50 kph constant speed

In any case, even with some approximations, this kind of contribution analysis can represent a starting point for the definition of quantitative directions for the reduction of the PB noise level of a given vehicle, as described in the next section.

4 A Procedure for PB Noise Target Setting

4.1 Description of the Principle – Unconstrained Case

In this section, results from Sects. 2 and 3 are combined to propose a sensible engineering procedure to cascade a target given in terms of PB noise level reduction onto targets for powertrain and tires noise reduction. Given the higher complexity of the ISO 362-1, such a target cascading process is not completely straightforward since the L_{urban} level is built by suitably "mixing" the levels of (up to) 4 driving conditions, for which the balance between the powertrain noise and the tires noise can be very different (see Figs. 6, 7, 8 and 9).

One assumes to be in the case in which the vehicle has to be tested in two different gear ratios. The modifications to be applied for the case in which only one gear ratio can be used will appear obvious.

Let us assume that a reduction of L_{urban} equal to a fixed ΔL_{urban} (in dB(A)) is desired. By looking at formula (8) *in a purely "formal" way*, it looks reasonable to "distribute" the required improvement among the different driving conditions in a way proportional to their weight coefficients in (8). In other words, in order to achieve the desired target, sub-targets for the 4 driving conditions are defined in such a way that

$$\Delta L_{urban} = k_P \cdot k_G \cdot \Delta L_{cru(i)} + k_P \cdot (1 - k_G) \cdot \Delta L_{cru(i+1)} + (1 - k_P) \cdot k_G \cdot \Delta L_{wot(i)} + \left(1 - k_p\right) \cdot (1 - k_G) \cdot \Delta L_{wot(i+1)} \tag{14}$$

$$\frac{\Delta L_{cru(i)}}{k_P \cdot k_G} = \frac{\Delta L_{cru(i+1)}}{k_P \cdot (1 - k_G)} = \frac{\Delta L_{wot(i)}}{(1 - k_P) \cdot k_G} = \frac{\Delta L_{wot(i+1)}}{(1 - k_P) \cdot (1 - k_G)} \tag{15}$$

From (14) and (15) it is possible to evaluate the ΔL required for the 4 different driving conditions in order to achieve the desired target.

Let us consider the example of the EU B-segment vehicle mentioned in previous sections and let $\Delta L_{urban} = 2$ dB(A). In this case, using the values of k_P and k_G reported in Sect. 2, Eqs. (14) and (15) give: $\Delta L_{wot(3)} = 2.87$ dB(A), $\Delta L_{cru(3)} = 0.87$ dB(A), $\Delta L_{wot(4)} = 1.08$ dB(A) and $\Delta L_{cru(4)} = 0.31$ dB(A). These values indicate that, as it appears fair from formula (9), for the achievement of the desired ΔL_{urban} target a reduction of the level measured in 3rd gear wot acceleration is required that is much higher than the reductions required for the other driving conditions. ΔLs required for 4th gear wot acceleration and 3rd gear cruise are about at the same level (around 1 dB(A)), while only a (relatively) minor improvement is required for 4th gear cruise conditions.

It is important to stress that all these operations are carried out in a rather formal way, simply looking at formula (8) and treating the different driving conditions as if they were completely independent one from the other which, of course, is not the case.

Once ΔL targets are available for each single driving condition, it is possible to cascade them onto targets for the reduction of powertrain and tires noise contributions.

This can be done using a procedure similar to the one explained in [16] and that is briefly recalled here below. For a given driving condition, the total SPL is the energetic sum of the powertrain and tires contributions:

$$SPL_{tot}(x) = 10 \cdot Log_{10}\left(\frac{p^2_{pwt}(x) + p^2_{tires}(x)}{p^2_{ref}}\right) \tag{16}$$

where p^2_{pwt} is the squared magnitude of the acoustic pressure generated by the powertrain in Pa^2 and p^2_{tires} is a similar quantity for the tires. From (16) it is possible to calculate the differential and obtain, after some simple algebra:

$$dSPL_{tot}(x) = \frac{p^2_{pwt}(x)}{p^2_{pwt}(x) + p^2_{tires}(x)} \cdot dSPL_{pwt}(x) + \frac{p^2_{tires}(x)}{p^2_{pwt}(x) + p^2_{tires}(x)} \cdot dSPL_{tires}(x) \tag{17}$$

which shows that

$$\frac{\partial SPL_{tot}}{\partial SPL_{pwt}} = \frac{p^2_{pwt}(x)}{p^2_{pwt}(x) + p^2_{tires}(x)} = \gamma_{pwt}(x) \tag{18}$$

$$\frac{\partial SPL_{tot}}{\partial SPL_{tires}} = \frac{p^2_{tires}(x)}{p^2_{pwt}(x) + p^2_{tires}(x)} = \gamma_{tires}(x) \tag{19}$$

where γ_{pwt} is the "fraction of noise" (in energetic terms) at the PB microphones that is due to the powertrain and γ_{tires} is a similar quantity for the tires. Formulas (18) and (19) basically state a very simple thing: that the sensitivity of SPL_{tot} to a variation of the powertrain (resp. tires) contribution in dB(A) is proportional to the fraction of SPL_{tot} (*in energetic terms*) that can be attributed to the powertrain (resp. tires). These formulas can be used to cascade the ΔL reduction required for a given driving condition onto ΔL_{pwt} and ΔL_{tires} reductions for powertrain and tires respectively, through a very fast and simple gradient-based optimization algorithm [16].

This was done for the test case of the previously mentioned EU B-segment vehicle, considering a target for PB noise reduction of 2 dB(A) and starting from the most demanding driving condition, namely the wot acceleration in 3rd gear. As mentioned previously, for this case $\Delta L_{cru(3)} = 2.87$ dB(A). Figure 10 shows the results of the target cascading process in terms of SPL values, while Fig. 11 shows the results in terms of ΔSPL for powertrain and tires. All these figures are limited to the range between −5 m and +10 m, where the maximum of the PB noise is localized. As one can see, the ΔSPL_{pwt} and the ΔSPL_{tires} required to achieve the desired 2.87 dB(A) reduction on the PB noise level are rather constant over this position range. Furthermore, given that in the baseline situation the contribution of the powertrain is higher than that of the tires, the ΔSPL reduction required for the powertrain is higher than the one required for the tires. Namely, ΔSPL_{pwt} is in the range of 3.3 dB(A) while ΔSPL_{tires} is in the range of 2.2 dB(A).

A similar procedure was applied to 3rd gear cruise, 4th gear wot acceleration and 4th gear cruise. Results obtained for the different driving conditions are shown in Figs. 12 and 13.

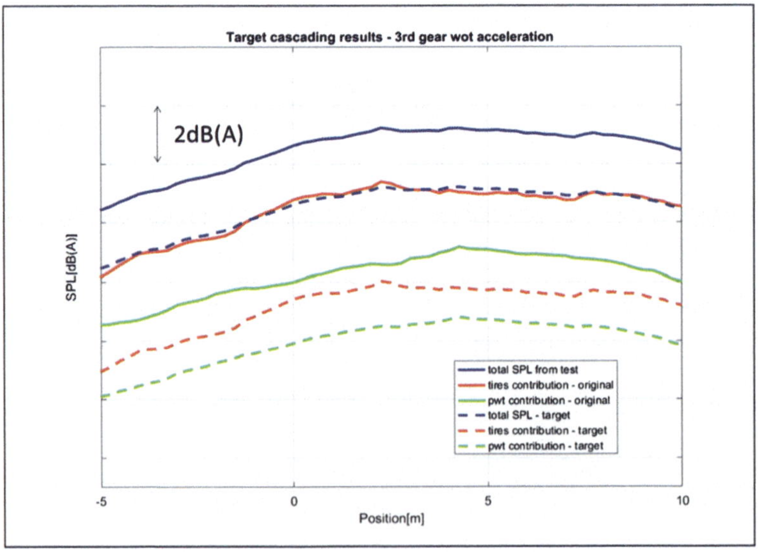

Fig. 10. Results of target cascading process for 3rd gear wot acceleration

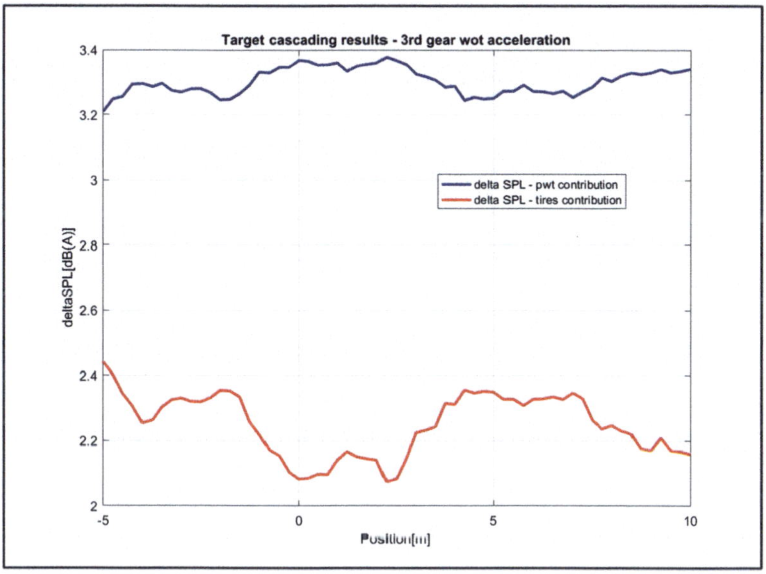

Fig. 11. Results of target cascading process for 3rd gear wot acceleration – SPL reductions required for powertrain (blue) and tires (red) contributions

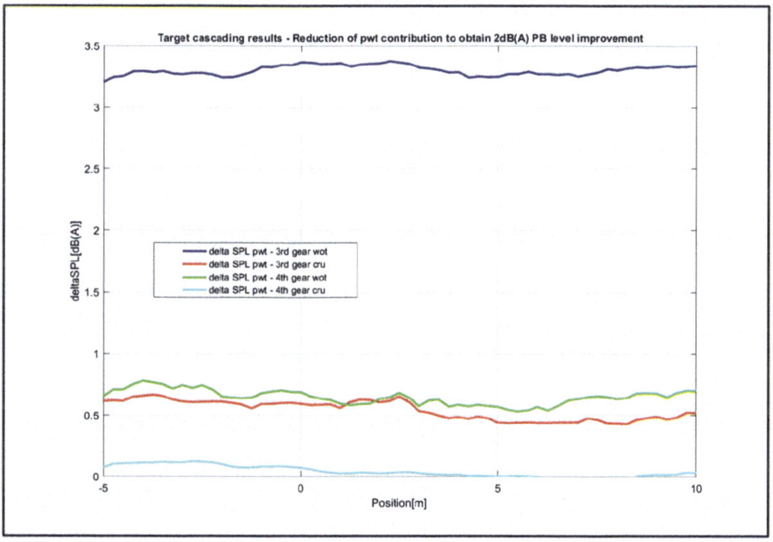

Fig. 12. Results of target cascading – SPL reduction required for powertrain contribution for different driving conditions

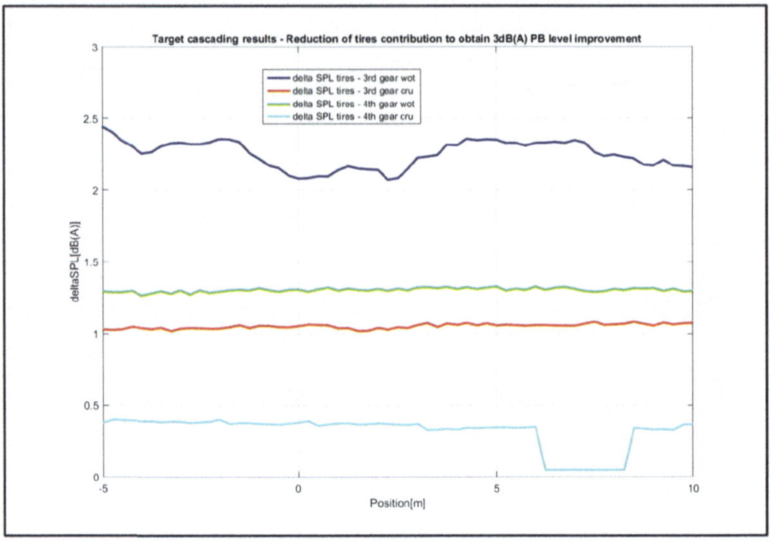

Fig. 13. Results of target cascading – SPL reduction required for tires contribution for different driving conditions

As one can see from Figs. 12 and 13, the reduction of powertrain contribution and tires contribution required to achieve the desired reduction of PB noise level depends in a rather strong way on the driving conditions. This makes sense, in relation to the different importance that the different driving conditions have in the definition of L_{urban} (see formula (8)).

As mentioned previously, in all this process the different driving conditions are treated in a totally independent way. In the end, though, it is not unreasonable to assume that if the ΔSPL reduction required for the "most demanding driving condition" (i.e. 3rd gear wot acceleration, for the test case considered here) is achieved, then this will happen also for the other driving conditions. This will be confirmed by the results reported in Sect. 5.

4.2 Application of Constraints

The target setting process described in previous section was done assuming that the powertrain and tires contributions to PB noise can be reduced without any limitation. Results in the technical literature [17] seem to indicate, though, that the reality is different. Limiting ourselves to noise reductions obtainable by means of passive treatments (i.e. not considering the possibility to work directly on the sources, tires or engine), it seems that with current technologies the reduction of the tires contribution to PB noise is limited to 1–1.5 dB(A). This is due to the fact that the main source of tire noise (in the frequency range 800–1.2 kHz, where the spectrum of tire noise has its maximum) is localized in the area of the tire/road contact [18–20], that cannot be effectively shielded with passive treatments. On the other hand, much higher reductions of the powertrain contribution seem feasible [21, 22], thanks to the possibility of designing treatments that "encapsulate" the engine and that can reduce its contribution to PB noise even by more than 5 dB(A).

The target setting process described in previous sub-section was then repeated by putting the following constraints:

- Maximum improvement allowed for tires contribution is 1 dB(A);
- Maximum improvement allowed for powertrain contribution is 6 dB(A).

Results obtained with these constraints are shown (in terms of ΔSPL_{pwt} and ΔSPL_{tires}) in Figs. 14 and 15.

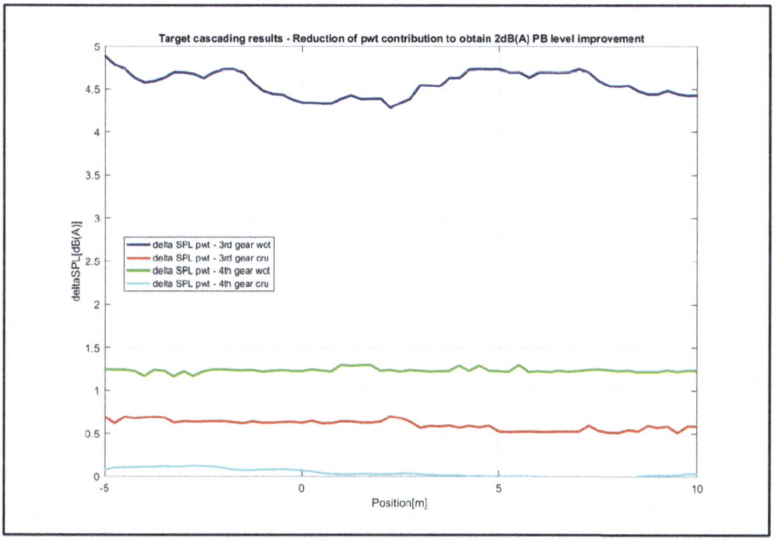

Fig. 14. Results of target cascading <u>with constraints</u> – SPL reduction required for powertrain contribution for different driving conditions

This latter figure shows that the required reduction of tires contribution reaches the maximum allowable for all driving conditions, with the exception of the constant speed in 4th gear, for which the improvement required at PB noise level is minimal. Correspondingly, the comparison of Fig. 14 with Fig. 12 shows that with the given constraints the improvement required for the powertrain contribution actually increases (as it had to be expected, of course) up to almost 5 dB(A).

Before concluding this section, results in Figs. 12, 13, 14 and 15 deserve a couple of further comments. The first comment is that, even though it is true that the new ISO 362-1 gives more importance to tire noise compared to the old ISO 362 measurement standard, in many cases the achievement of a given PB level target still relies mainly on the reduction of powertrain noise. This is due in good part to currently existing constraints on the noise reduction obtainable for tires noise but it is due also

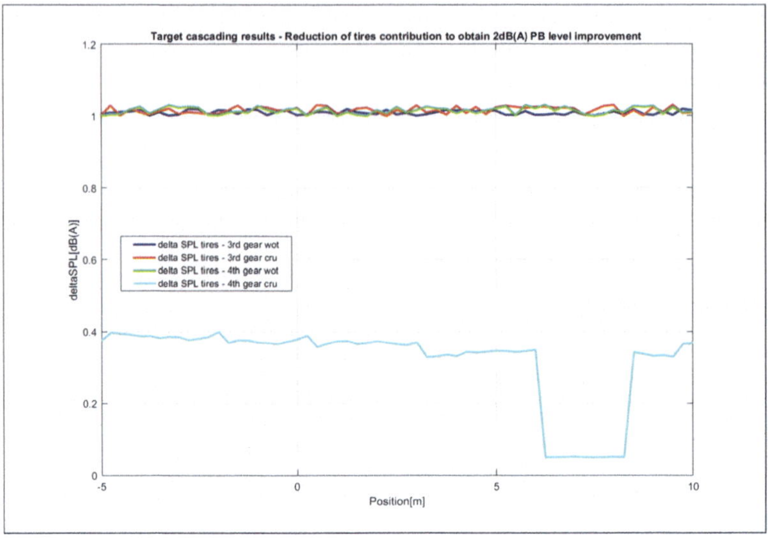

Fig. 15. Results of target cascading <u>with constraints</u> – SPL reduction required for tires contribution for different driving conditions

to the weight that wot-accelerations have in the "composition" of L_{urban} (see formulas (9) and (11)), that is substantially bigger than that of constant speeds. The second comment is about the fact that the data appearing in Figs. 14 and 15 basically show that, at least in the case considered here, reaching the R51.03 Phase 3 target is really pushing the reductions obtainable with passive noise treatments to their limit. This situation is somehow intuitively known to NVH engineers working on exterior noise since some time (see, e.g. [23], published in 2006). The merit of the procedure proposed in this section is to get to this result in a clear and analytical way and in giving the possibility to update and re-balance the noise reductions required for tire and engine on the basis of the "boundary conditions" available, i.e. the specificities of the vehicle considered, the design constraints, etc.

5 PB Noise Target Achievement – An Application Example

Once targets for the reduction of powertrain and tires contributions to PB noise are given, suitable sound package solutions must be found to achieve them. This section describes an example of such "target achievement process", essentially recalling results reported in [5] and giving more details about some aspects of the activities that were carried out.

Starting point for all the activity was the need to reduce the PB noise level by 2 dB(A). Indications given by the OEM for reduction of powertrain and tires noise were the following: "…an expectation of −5 dB(A) on the engine noise contribution and −0.5 dB (possibly even −1.0 dB) on the tire noise contribution" [5]. These expectations were developed by the OEM on the basis of outdoor tests, but they match well with the results reported on previous section.

5.1 Definition of the Sound Package "Design Space"

The first step of the activity consisted in defining a kind of "sound package design space", i.e. put on the table all feasible sound package options for the reduction of powertrain and tires exterior noise and then realize the corresponding prototypes. Figures 16, 17 and 18 show the treatments that were considered for the engine bay, the underbody and the heat-shields areas respectively. Once this "design space" is defined, the next problem consists in exploring it, i.e. in understanding what is the effect of the different sets of parts on the PB noise level and how to best combine them in order to reach an optimal weight/cost/performance ratio. For this, a test matrix with about 25 different configurations was defined, combining the treatments in several different possible ways. Given that the assessment of all these configurations by means of outdoor measurements on a test-track was out of question due to the time and resources required, it was decided to make a first analysis of the test matrix with indoor tests, as described in the next sub-sections.

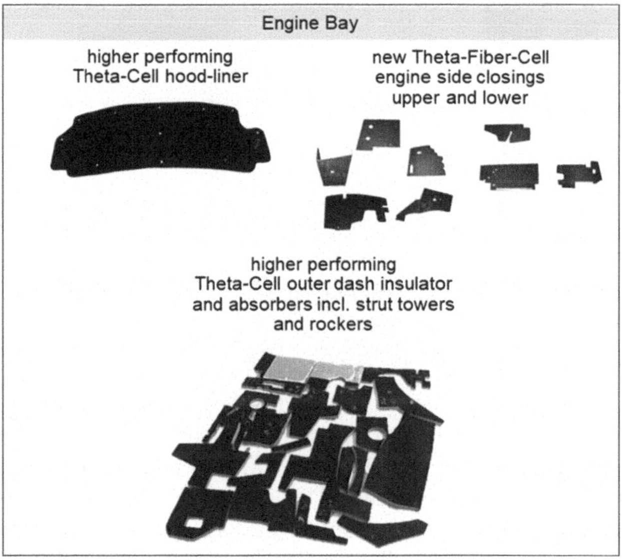

Fig. 16. Possible engine bay treatments considered for improvement of PB noise

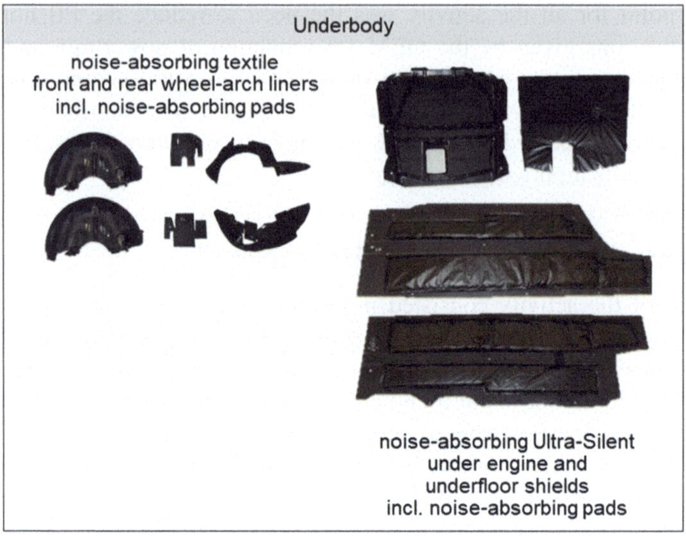

Fig. 17. Possible underbody treatments considered for improvement of PB noise

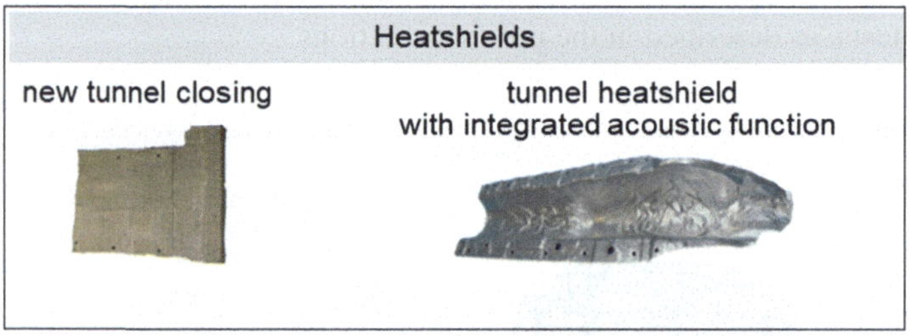

Fig. 18. Possible heat shield treatments considered for improvement of PB noise

5.2 Analysis of Sound Package Impact on ATFs Relative to PB Microphones

As a first step, Acoustic Transfer Functions (ATFs) were measured for all 25 configurations included in the test-matrix. Figure 19 shows the topological layout of the experimental set-up used. PB microphones were distributed between $x = -2$ m and $x = 6$ m on the right side of the vehicle, since this was known to be the position range where the maximum in PB noise occurs. The analyses of the results obtained allowed ranking the different combinations of sound package parts investigated, in terms of the impact they have on the ATFs between the engine and the PB microphones and between the tires and the PB microphones.

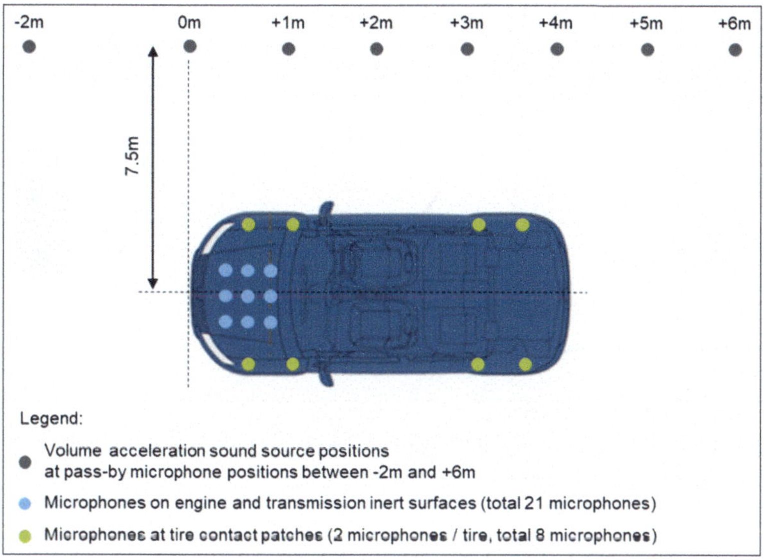

Fig. 19. Experimental lay-out for measurement of engine-to-PB-microphones and tires-to-PB-microphones ATFs

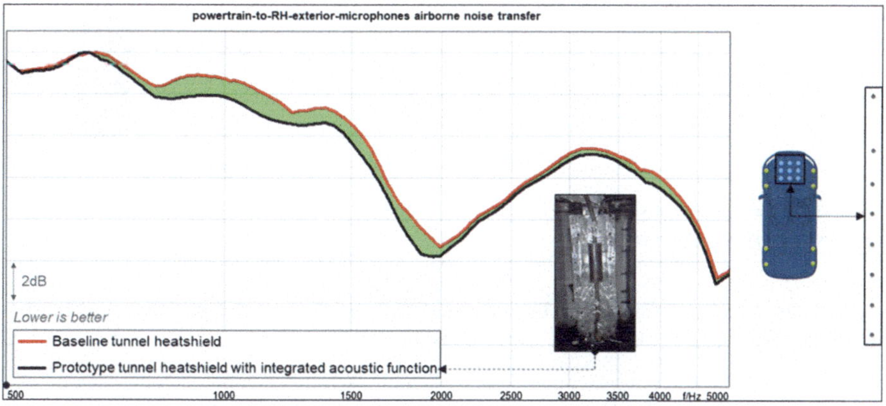

Fig. 20. Effect of acoustic heatshield on engine-to-PB-microphones ATF – delta ATF represented in green indicates an improvement

Just as an example, Fig. 20 shows the impact on the engine ATF of the acoustic tunnel insulator displayed in Fig. 18. At the end of all the activity on the ATFs, a feasible package (hereafter called "PB noise package" or simply "PB package") was selected, able to reduce the engine bay ATF by 4–5 dB(A) over most of the frequency range (see Fig. 21) and able to reduce the tires ATF by about 2 dB(A) in the critical frequency range around 1 kHz (see Fig. 22). More details about the sound package configurations analyzed in this phase and the package finally selected can be found in [5].

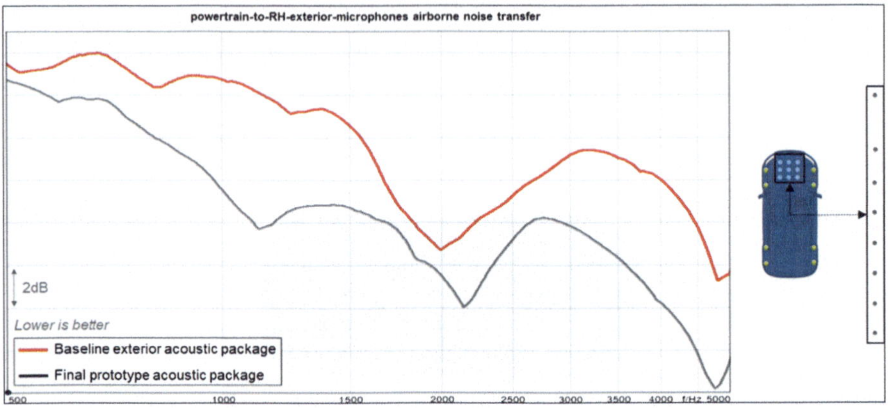

Fig. 21. Engine-to-PB-microphones ATF measured with original sound package (red line) and with "PB noise package" (black-line). Lower is better.

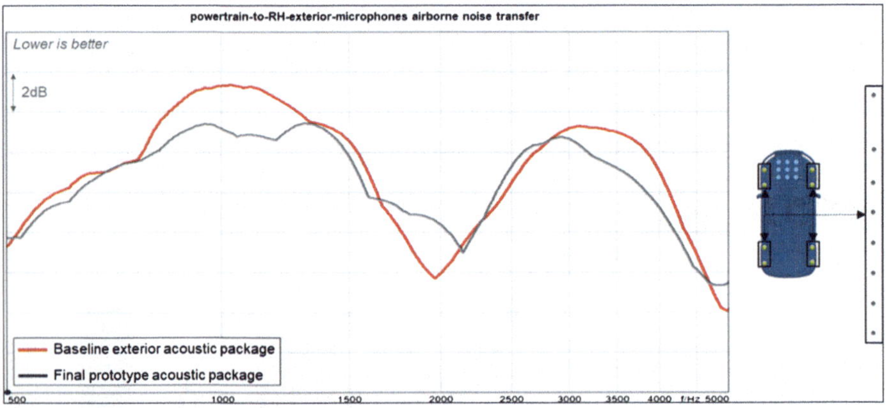

Fig. 22. Tires-to-PB-microphones ATF measured with original sound package (red line) and with "PB noise package" (black-line). Lower is better.

5.3 Indoor PB Noise Tests

Indoor PB measurements offer some advantages over outdoor tests: easier to perform in a very controlled way and thus more repeatable, independent from weather conditions, faster. For this reason, in the last years there was a relevant push towards some kind of standardization of these measurements, that led to the definition of the ISO 362-3 norm [11, 12]. Still, they suffer from the fact that at the moment it appears difficult to represent adequately through indoor tests the contribution to exterior noise given by the tires [13]. For this reason, ISO 362-3 suggests to use indoor tests only for the estimation of the powertrain contribution and to resort to outdoor tests for the estimation of the tires contribution. An exercise following the guidelines of ISO 362-3 was carried out in order to validate the "PB noise package" before trying it outdoor on the test track.

More specifically, the following procedure was followed:

a. First of all, it was verified that *in baseline conditions* it was possible to get a good correlation between the outdoor and the indoor PB noise tests;
b. Secondly, indoor PB noise tests were carried out with the "PB noise package" installed and the impact on the overall PB noise level was estimated, as well as the impact on the powertrain contribution and on the tires contribution to such level.

As far as step A is concerned, standard indoor PB tests were carried out as well as coast-down tests to allow the evaluation of the FRN, in a way totally similar to what was described in Sect. 3 for outdoor tests. After this, indoor measurements were corrected by first subtracting (energetically) the indoor-measured FRN and then adding (energetically) the outdoor-measured FRN. This way of correcting the indoor measurements led to the results shown in Fig. 23 for 3rd gear wot and in Fig. 24 for 3rd gear cruise (similar results are obtained for 4th gear measurements). As one can see from these figures, once the FRN correction is taken into account, the correlation between indoor and outdoor tests is more than acceptable, as reported also in [11].

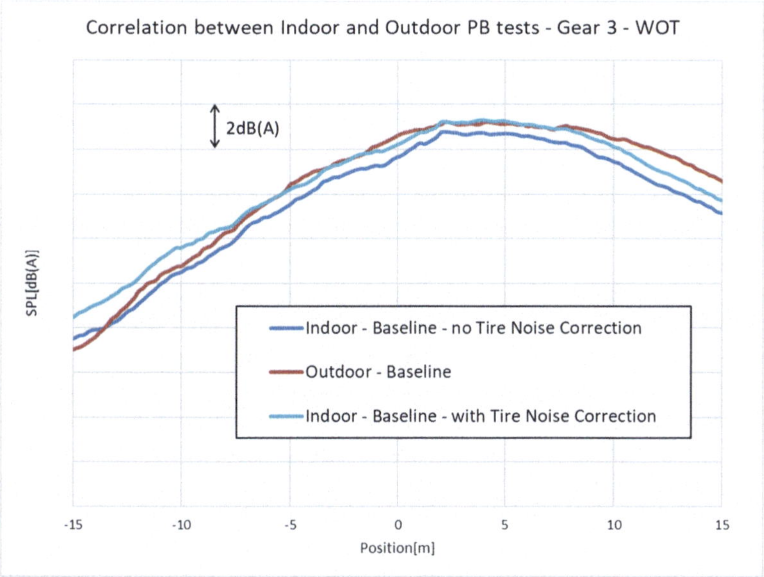

Fig. 23. Comparison between indoor and outdoor tests in baseline conditions – 3rd gear wot acceleration

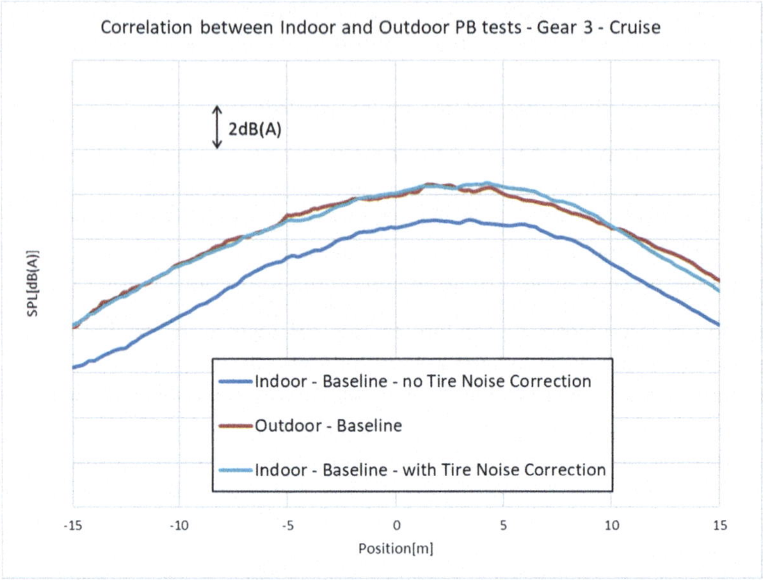

Fig. 24. Comparison between indoor and outdoor tests in baseline conditions – 3rd gear 50 kph constant speed

After checking that results from indoor tests could well represent (in terms of overall level) results from outdoor tests, next step consisted in installing the "PB noise sound package" on the test vehicle and verify its impact both on the PB noise level and on the engine and tire noise contribution.

Figures 25, 26, 27 and 28 show the impact of the application of the "PB package" on the indoor-measured PB noise for all analysed driving conditions. As one can see, this impact is very relevant, in particular for wot accelerations in 3rd and in 4th gear. The impact on *indoor* L_{urban}, calculated according to ISO 362-1, turned out to be about **3.5 dB(A)** (see [5]), which is even higher than initially expected (as previously explained, an improvement of about 2 dB(A) on L_{urban} was targeted).

To get some more insight on the reduction determined by the applied sound package on the powertrain and tires contributions, a Source Contribution Analysis (SCA) was carried out. For this SCA, the method described in [6, 10] was used. The method is based the so called "Substitution Monopole Technique" (SMT, [14, 15]) on the idea that the radiation from the relevant noise sources of the car (tires, engine, etc.) can be modelled by means of a set of monopoles, suitably distributed in the vicinity of the surface of these same sources. The number and the position of these "fictitious monopoles" is defined by the user on the basis of experience and on the basis of some knowledge of the physical mechanisms that govern the noise radiation from a certain source.

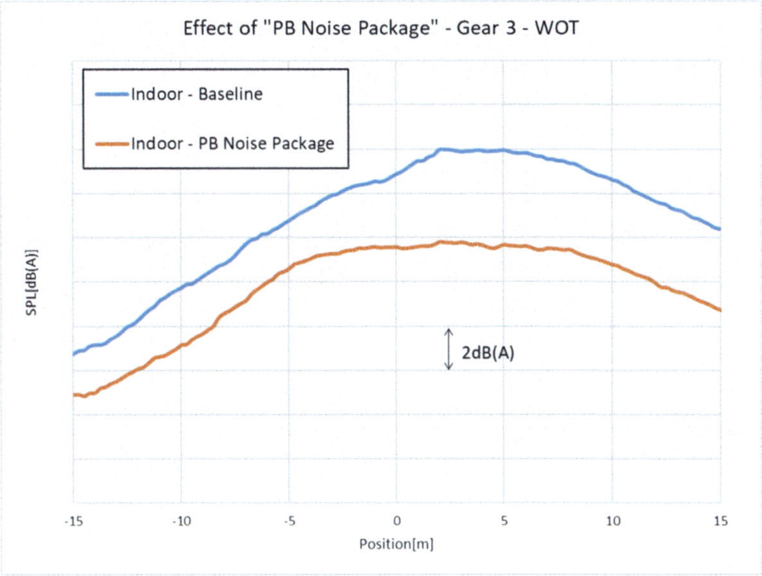

Fig. 25. Impact of "PB package" on indoor PB noise – 3rd gear wot acceleration

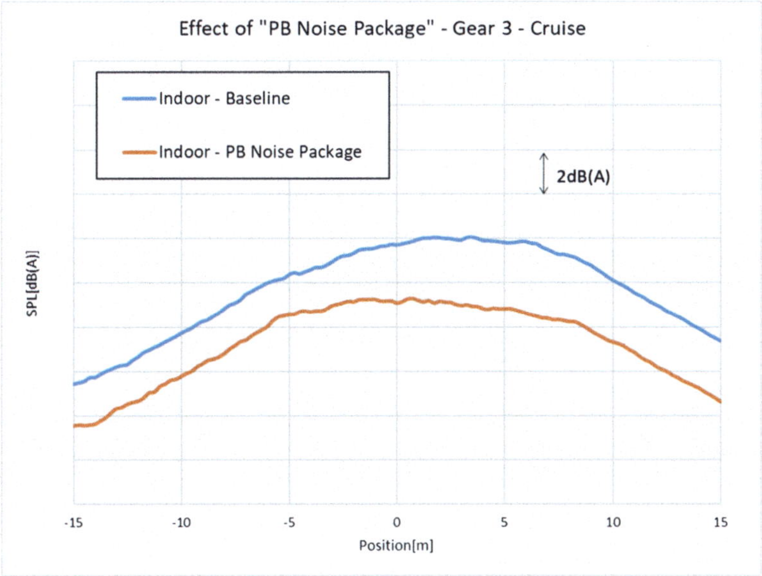

Fig. 26. Impact of "PB package" on indoor PB noise – 3rd gear 50 kph constant speed

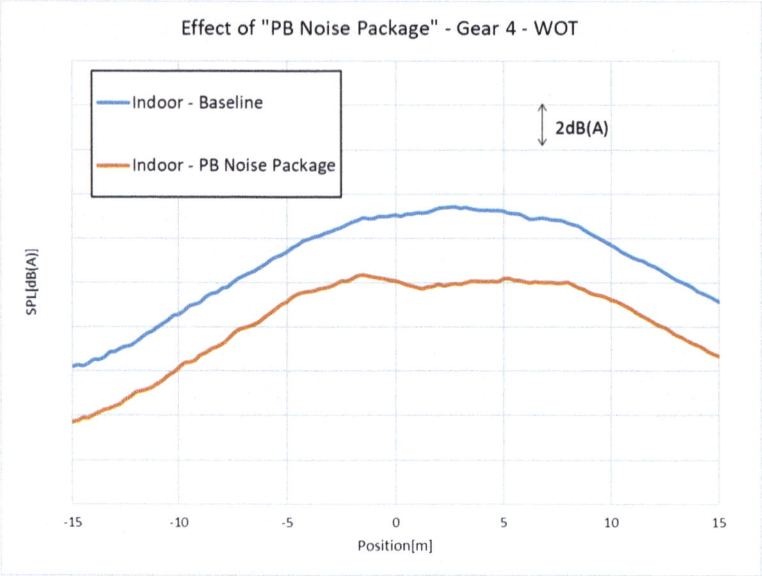

Fig. 27. Impact of "PB package" on indoor PB noise – 4th gear wot acceleration

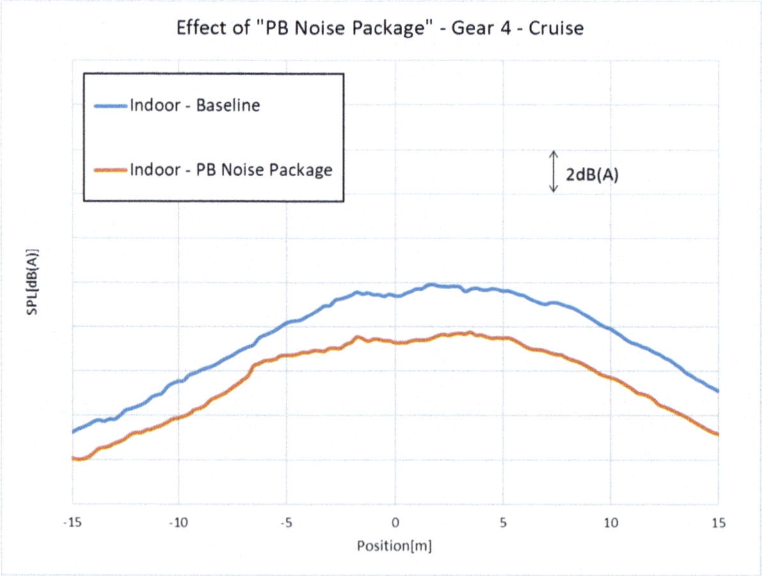

Fig. 28. Impact of "PB package" on indoor PB noise – 4th gear 50 kph constant speed

Once the position of these monopoles is decided, the next problem consists in defining the corresponding source strengths for the different driving conditions analysed. This is done in the following way: the SPL at a set of "indicator microphones" around the source is measured and, after this, the p/Q Acoustic Transfer Functions (ATFs) between the source positions and the indicator microphones are evaluated. By pseudo-inverting the ATFs matrix it is then possible to estimate the source strengths of the monopoles that somehow "model" in the best possible way the source, on the basis of the data measured at the indicator microphones (see Fig. 29).

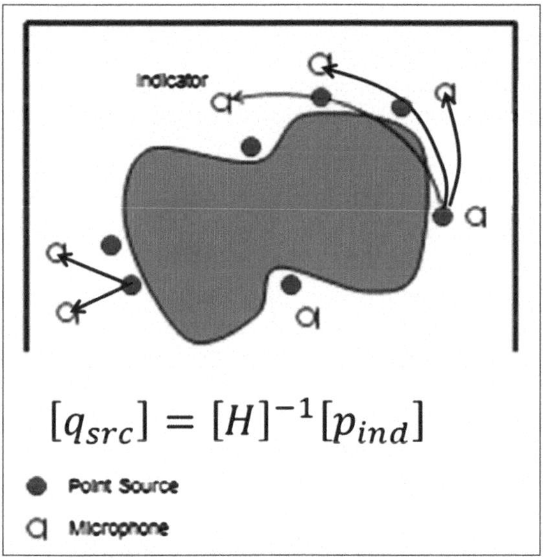

Fig. 29. Evaluation of source strengths from operative levels at indicator microphones and transfer functions (figure taken from [6])

Once the strengths of these monopoles are given, their contribution to a certain receiver microphone can be obtained by multiplying these strengths by the ATFs between the monopoles positions and the receiver positions (i.e. the PB microphones, in our case). The contribution of a physical source is obtained simply by summing up the contributions of all the corresponding monopoles (see Fig. 30).

For the case of the B-segment vehicle considered here three different "contributions" were analyses:

- The powertrain, modelled with twelve monopoles distributed over its surface (two per face);
- The tyres, modelled each with two monopoles (one near the leading edge and the other near the trailing edge of the contact patch);
- The intake, modelled with one monopole positioned at the intake mouth.

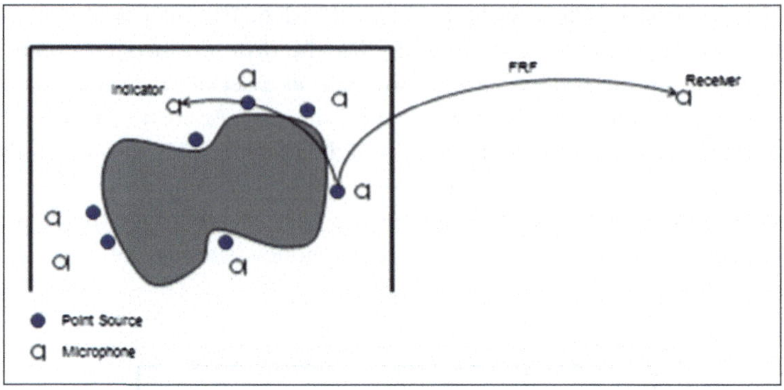

Fig. 30. Evaluation of source contributions from source strengths and ATFs between source positions and receiver positions (i.e. PB microphones). Figure taken from [6].

During indoor tests, the exhaust contribution was excluded (by the application of a big silencing muffler) since previous measurements by the car maker indicated that it was in the order of 5%, thus negligible [5].

Figure 31 shows -just as an example- the result of the contribution analysis for a 3rd gear wot acceleration for the vehicle in baseline conditions. Results for the other driving conditions are analogous.

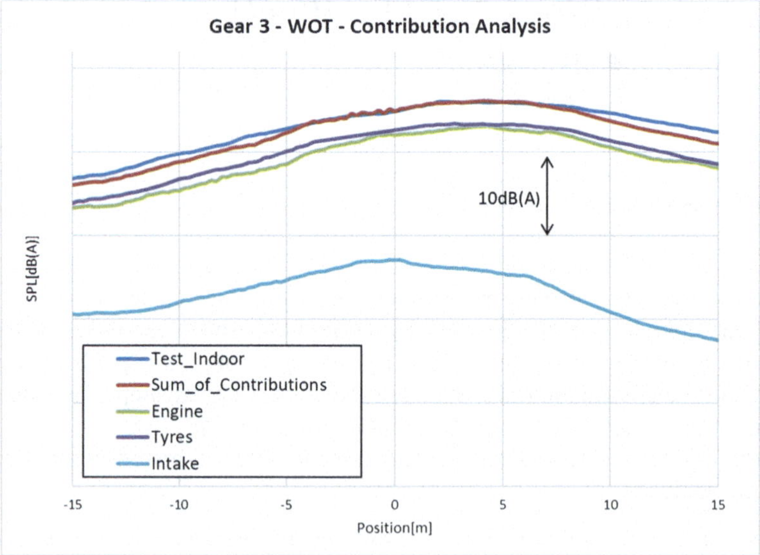

Fig. 31. Results of Source Contribution Analysis from indoor PB noise tests – 3rd gear wot acceleration

From this figures it appears that the "sum of the contributions" is generally able to reproduce within 1–1.5 dB(A) the PB noise level, in particular in the positions range around the maximum. This is a good indication that no relevant contribution to the PB noise was missed in the analyses. Furthermore, it is also clear that the intake contribution can, in general, be considered negligible. It is also interesting to notice that, even in 3rd gear wot acceleration, the contribution from the tires is comparable to the one of the engine. Comparing results of Fig. 31 with results in Fig. 6 it appears clear that, even in qualitative terms, the relative contribution from the tires is likely to be higher when measured indoor than when measured outdoor.

This SCA exercise based on SMT was repeated both for the vehicle in baseline conditions and for the vehicle equipped with the "PB package". At the end, by difference the impact of this same package on the powertrain and on the tires contribution could be evaluated, for all driving conditions. Figure 32 shows the reduction of the powertrain contribution to PB noise obtained in this way. Similar results are shown in Fig. 33 for the tires noise contribution. As one can see from these figures, the reduction of powertrain contribution is -in the range of the SPL maximum, i.e. between x = 0 m and x = 5 m- around 4–5 dB(A) as required by the target setting procedure described in previous sections. It is also interesting to observe that such reduction does not depend too much on the driving conditions. On the other hand, the reduction of the tires contribution is more dependent on the driving conditions and, furthermore, it is generally higher than expected. This is likely to be due to the fact that the indoor-measured tire noise has a high-frequency spectral content that is (for the same overall level) more pronounced than that of the outdoor-measured tire

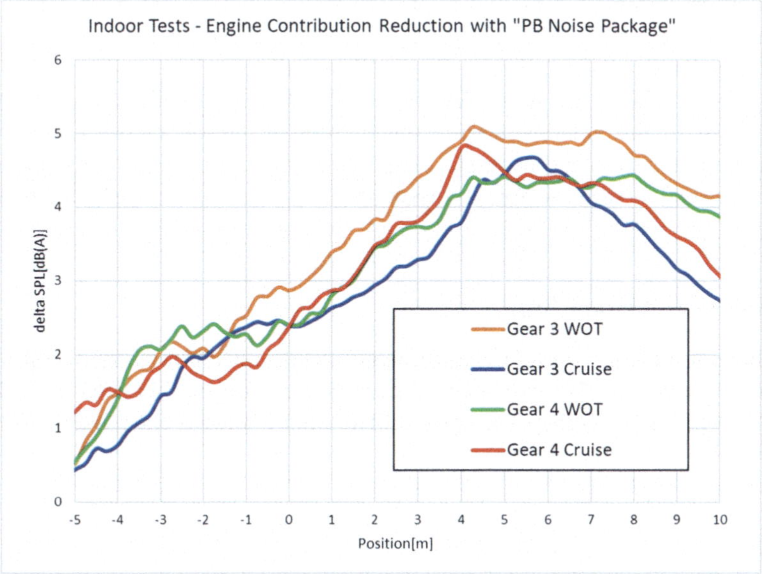

Fig. 32. Reduction of engine contribution to PB noise obtained with "PB Package" – estimation with Substitution Monopole Technique

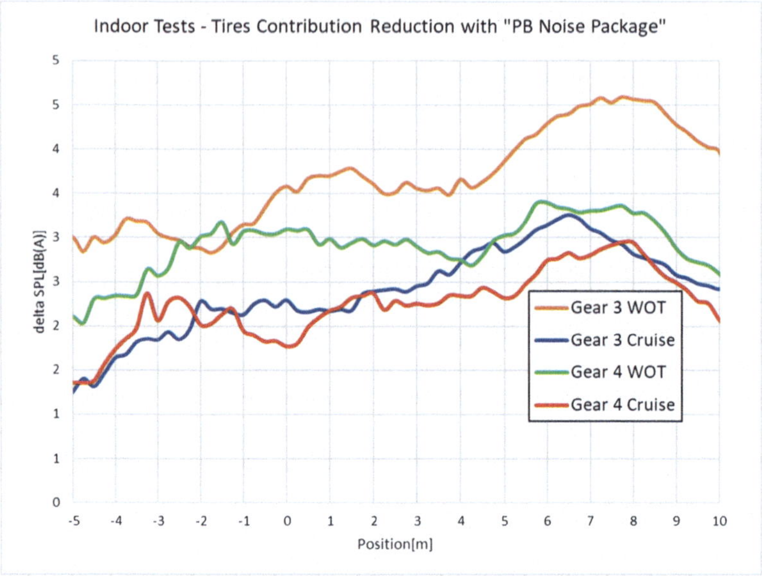

Fig. 33. Reduction of tires contribution to PB noise obtained with "PB Package" – estimation with Substitution Monopole Technique

noise [11]. Most of the solutions implemented in the "PB package" for tire noise (e.g. absorptive wheel-house outer liners or underbody shields) are of an absorptive type and their effectiveness is thus higher at high frequencies. Combining these 2 facts, it is clear that indoor PB tests might lead to some overestimation of the impact that these solutions have on tires contribution to PB noise.

This shows, once again, that indoor tests should be used with care when it comes to the analysis of tire noise, since the specificities of the indoor tire noise spectrum can somehow bias even the estimate of the effectiveness of a certain package. [24] describes a possible solution for this issue that seems to be, though, rather resource consuming.

5.4 Outdoor Validation Tests

Final validation of the proposed "PB package" was carried out by the OEM on a test track. Results (see Fig. 34) indicated an improvement on the final PB noise level of 2 dB(A) [5], in line with the expectations. The final PB noise level obtained fulfilled the R51.03 Phase3 limit.

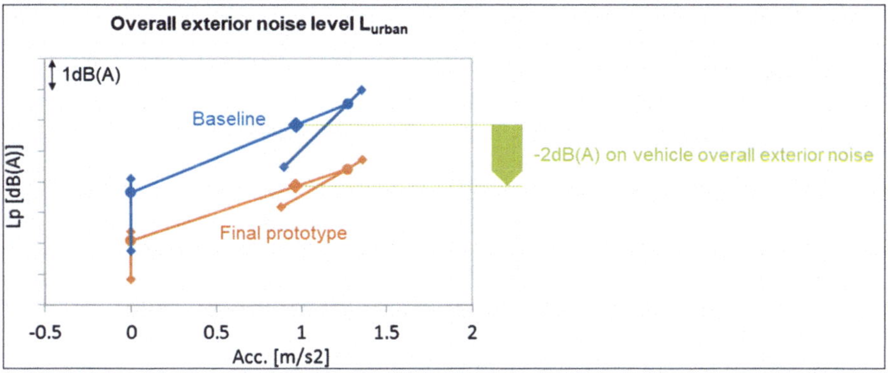

Fig. 34. Results of final outdoor validation of "PB noise package" by OEM

6 Conclusion

This paper presented an engineering procedure for the definition of a road-map to improve the PB noise level of a vehicle. This procedure combines a simple analysis of the way the level L_{urban} defined by the ISO 362-1 is "built up" (starting from the levels measured in different driving conditions) with a more challenging analysis of the contributions that the powertrain and the tires give to these same levels.

Like any other procedure, it cannot (and it should not) be applied in a blind way, but it can give some useful hint about how to make the achievement of the desired PB noise target more efficient, reducing the need for trial-and-error experimental campaigns.

The paper presented also an application of this procedure to a B-segment EU car, where the PB noise level was reduced below the threshold defined by the R51.03 Phase 3, thanks to the application of a suitable sound package. The extensive work carried on this vehicle was a combination of indoor and outdoor tests. Indoor tests turned out to be very useful to have the possibility to analyse several different sound package solutions and, in the end, sort out the most promising one. At the same time, all the work carried out showed also that indoor tests, while being very useful, still have strong limitations for what concerns in particular the analysis of tire noise and the effect that sound packages can have on it.

References

1. Moore, D., "Development of ECE R51.03 Noise Emission Regulation," SAE Technical Paper 2017-01-1893, 2017, https://doi.org/10.4271/2017-01-1893
2. ISO 362-1:2015 standard. Measurement of noise emitted by accelerating road vehicles – Engineering method – Part 1: M and N categories

3. Steven H. Investigations on Improving the Method of Noise Measurement for Powered Vehicles, Report Number 10506067 by order of the Germany Federal Environmental Agency, August 1999
4. Steven H. Further Noise Reductions for Motorized Road Vehicles. Presentation within the Workshop of the German Federal Environmental Agency, September 2001
5. Y. Denoual et al, "Vehicle NVH Pack Optimization for Exterior Noise using Indoor Experimental Testing", Proceedings of SIA 2018, LeMans, 2018
6. A. Schumacher, E. Shirahashi, "Indoor pass-by noise contribution analysis using source path contribution concept", Proceedings of 2012 ISMA Conference, Leuven, 2014
7. J. Putner, M. Lohrmann, and H. Fastl, "Contribution analysis of vehicle exterior noise with operational transfer path analysis", Proceedings of ICA 2013, Montreal, 2013
8. K. Janssens, F. Bianciardi et al, "Pass-by noise engineering: a review of different transfer path analysis techniques", Proceedings of ISMA 2014, Leuven, 2014
9. K. Janssens, P. Aarnoutse, "Time-Domain Source Contribution Analysis Method for In-Room Pass-By Noise", SAE Technical Paper 2011-01-1609, 2011
10. Z. Chou et al, "Source Path Contribution Analysis for Vehicle Indoor Pass-By Noise", Sound and Vibration Magazine, September 2017
11. Krämer M., Brandstätt P., Ickinger A., "Comparison of pass-by noise from real track and simulated measurement at the roller test-bench", Proceedings of Internoise 2016, Hamburg, 2016.
12. ISO 362-3:2016 standard. Measurement of noise emitted by accelerating road vehicles – Engineering method – Part 3: Indoor testing M and N categories
13. Krämer M., Brandstätt P., Bošković B. (2014) Comparison of vehicle rolling noise on the test bench and on the road. In: Bargende M., Reuss HC., Wiedemann J. (eds) 14. Internationales Stuttgarter Symposium. Proceedings. Springer Vieweg, Wiesbaden
14. J.W. Verheij, "Reciprocity method for quantification of airborne sound transfer from machinery", Second international congress on air- and structure-borne sound and vibration", Auburn University, USA, March 4–6, 1992
15. J.W. Verheij, A.N.J. Hoebrichts, D.J. Thompson, "Acoustical source strength characterisation for heavy road vehicle engines in connection with pass-by noise", Third international congress on air- and structure-borne sound and vibration", Montreal, Canada, June 13–15, 1994
16. C. Bertolini, "An SEA-based Procedure for the Optimal Definition of the Balance between Absorption and Insulation of Lightweight Sound Package Parts", SAE Technical Paper 2012-01-1527, 2012
17. T. Lafont, J. Horak, R. D'Amico, R. Stelzer and C.Bertolini. Passive treatment solutions for the reduction of vehicle exterior tire noise. 10th International Styrian Noise, Vibration & Harshness Congress, June 20–22, Graz, 2018
18. W. Kropp, F.-X. Bécot and S. Barrelet, On the Sound Radiation from Tyres, Acustica – Acta Acustica, vol 86(5), 769–779, 2000
19. W. Kropp, P. Sabiniarz, H. Brick and T. Beckenbauer. On the sound radiation of a rolling tyre. J. Sound Vib., 331, 1789–1805, 2012
20. T. Lafont, W. Kropp et al. – Modeling Tyre noise in FE simulations for pass-by noise predictions, ARCHIVE Proceedings of the Institution of Mechanical Engineers Part C Journal of Mechanical Engineering Science 1989–1996 (vols 203–210) https://doi.org/10.1177/0954406219832908
21. C. Bertolini, D. Caprioli, J. Horak, T. Bürgin, D. Petley, W. Jansen, B. Wicksteed, Acoustic benchmark of different CO_2 encapsulation strategies: application examples and methods for the design of powertrain encapsulation, Automotive Acoustic Conference, Zürich (2013)

22. R. D'Amico, R. Stelzer, A. Bihhadi, T. Courtois, Engine encapsulation: impact on vehicle NVH and CAE challenges, International Conference on Noise and Vibration Engineering (ISMA2016), Leuven (2016)
23. V. Rathje, "Lärmarme PKV und ihr Einfluss auf den Verkehrslärm – Stand der Technik und künftige Minderungspotentiale», II. Internationale Lärmschutzkonferenz "Umsetzung der EG-Umgebungslärmrichtlinie", Celle, 2006
24. Goto, K., Kondo, T., Takahira, M., Umemura, E. et al., "Indoor Pass-by Noise Evaluation System Capable of Reproducing ISO Actual Road Surface Tire Noise," SAE Technical Paper 2016-01-0479, 2016, https://doi.org/10.4271/2016-01-0479

Utilizing a Source-Path-Contribution Approach with a Desktop NVH Simulator to Improve Powertrain NVH Development Efficiency

Chris Cundy(✉), Mark Stickler, and Richard Luken

Ford Motor Company, 2400 Village Rd, Dearborn, MI 48121, USA
ccundy1@ford.com

Abstract. Evaluating powertrain sounds in a virtual vehicle environment is a key technology enabler for improving NVH development efficiency. NVH Simulator models constructed utilizing a source-path-contribution approach allow for assessment of individual subsystem contributions during simulated drives, all in the absence of prototype vehicles. Application examples are presented to show how the tool is utilized to facilitate powertrain NVH hardware/target compatibility decisions in the early stages of program development.

Keywords: Powertrain · NVH · Sound · Simulator · Interior · TPA

1 Introduction

1.1 Advantages of Desktop Simulation Modeling Over Traditional In-Vehicle Powertrain Sound Evaluation

Traditional in-vehicle powertrain sound evaluation has several limitations that inhibit its usefulness for assessing NVH targets and hardware compatibility. First, vehicle prototypes are typically unavailable in the early development stages of a new program when key powertrain hardware and NVH target decisions are being made. Further, due to the very high cost associated with building prototypes, automotive OEMs are working to reduce prototype testing or altogether eliminating certain prototype build phases. Even when suitable prototypes or surrogate vehicles are available, performing in-vehicle evaluations is time-consuming and expensive. Often it is impractical to evaluate alternative subsystem or component designs on a vehicle due to excessive cost, limitations of hardware prototyping, or packaging/installation incompatibility with the donor vehicle. Back-to-back in-vehicle comparisons are also hindered by time between evaluations (reliance on auditory memory), and variations in driving behavior among evaluators may lead to different experiences and conclusions. These limitations can be overcome by evaluating powertrain sound using a desktop vehicle NVH simulator.

W. Siebenpfeiffer (Hrsg.): *Automotive Acoustics Conference 2019*, Proceedings, S. 170–184, 2020. https://doi.org/10.1007/978-3-658-27669-0_12

A desktop vehicle NVH simulator presents sounds in a virtual vehicle environment, allowing for much of the interactivity and context one would experience during an actual vehicle drive evaluation [1]. For drivable models the evaluator is able to use pedal and steering wheel inputs to control the vehicle as it traverses a virtual environment. Vehicle noise representative of the driving conditions is synthesized and played through headphones. The Brüel & Kjær (B&K) Desktop Simulator (DTS) system is shown in Fig. 1.

Fig. 1. B&K Desktop Simulator (DTS) system

Desktop simulator models may be constructed well before prototype vehicles are available using measurement data from a surrogate vehicle or powertrain system, and data from subsystem-level hardware prototype measurements and CAE analyses can be incorporated to aid in program content and target decisions. The simulator environment allows evaluators to toggle back and forth instantaneously among multiple models/configurations to highlight differences and reduce reliance on auditory memory. Further, since several evaluators may simultaneously experience the same driving maneuver and associated auditory and visual stimuli, uncertainty introduced by inconsistent driving maneuvers and other factors that can affect powertrain/vehicle behavior can be eliminated.

1.2 Integration of Transfer Path Analysis Approach with Desktop Simulation

Transfer Path Analysis (TPA) models utilizing a source-path-contribution (SPC) approach are often employed in the early stages of new vehicle program development to cascade NVH targets to individual powertrain subsystems and to assess total interior powertrain noise status relative to a program target. Using the SPC approach, interior noise contributions from the important powertrain subsystems are computed based on a source-level characterization (e.g. exhaust tailpipe noise near the tailpipe orifice) and a transfer function (e.g. the acoustic transfer function between tailpipe orifice noise measurement location and driver's outboard ear). The total interior noise can be represented by the sum of the SPC contributions in addition to other direct contributions, e.g. wind noise or interior lugging boom predictions from a CAE model, according to Eq. 1.

$$Interior\ Noise = \sum \left[(source) * (transfer\ function)\right] + Remainder \tag{1}$$

In cases where an incompatibility between powertrain hardware content and an NVH target is identified, it can be difficult to align program decision makers around a resolution based solely on numerical or graphical data. The customer impact often needs to be better understood to help determine if the target deviation should be allowed or if NVH performance needs to be improved via a hardware (or calibration) change that may increase cost or entail trade-offs with other attributes. By presenting the predicted interior sound in the desktop simulator environment with interactivity and context, it is much easier for NVH experts and non-experts alike to assess the customer impact and align on how to proceed.

The B&K DTS environment facilitates building models using the SPC approach, which allows for subjective evaluation of various powertrain hardware designs simply by changing the associated source-level data. Likewise, design actions affecting vehicle powertrain structureborne or airborne transfer functions as well as road and wind noise can be evaluated. Comparator vehicle models (typically based on direct interior noise measurement rather than the SPC approach) such as the outgoing product or a competitive product can also be included in evaluations.

1.3 Use of Desktop Simulator Models with SPC Approach in Product Development Process

The desktop simulator can aid the powertrain NVH development process in various stages of the product development cycle (Fig. 2). In the pre-program phase when the high-level assumptions are being defined it may be used to help with powertrain unit selection, e.g. comparing an inline 4-cylinder (I4) engine with a V6 engine in a new vehicle program to evaluate sound character and performance to NVH targets, or to help with major powertrain NVH subsystem architectural decisions, e.g. determining if switchable or active engine mounts are required to meet interior loudness and other attribute targets. As program assumptions become more clearly defined, simulator studies can be utilized to help demonstrate the specific powertrain hardware content/design required to meet NVH targets and align the team around any contested content requests. Later in program development as system and sub-system level performance is quantified through CAE and/or measurement, the simulator can be used to help align the team on how to disposition target misses and evaluate potential solutions. The ability to make these decisions in a virtual vehicle environment is a key enabler for reduction or elimination of prototypes.

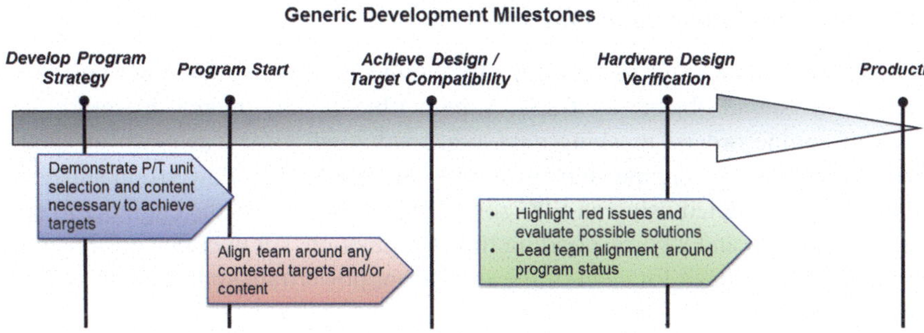

Fig. 2. Desktop simulator usage in the powertrain NVH development process

2 Technical Approach

2.1 Desktop Simulator Model Hierarchy

The hierarchy of a powertrain NVH DTS model using the SPC approach is shown in Fig. 3. Typically, the four most significant powertrain noise contributors are represented: powerplant (engine and transmission assembly) radiated noise, powerplant mounts structureborne noise, intake orifice noise, and exhaust orifice noise. DTS models are scalable, so additional contributions such as exhaust muffler shell radiated noise or driveline structureborne noise may be included depending on the specific study to be conducted. Road, wind, and tire noise contributions are also included. A vehicle performance model or pre-defined driving maneuver must also be specified to govern the sound playback and progression of visual elements (virtual road and surroundings, speedometer, tachometer, transmission gear, etc.) within DTS. The process for defining each of these model components is described in the following sections.

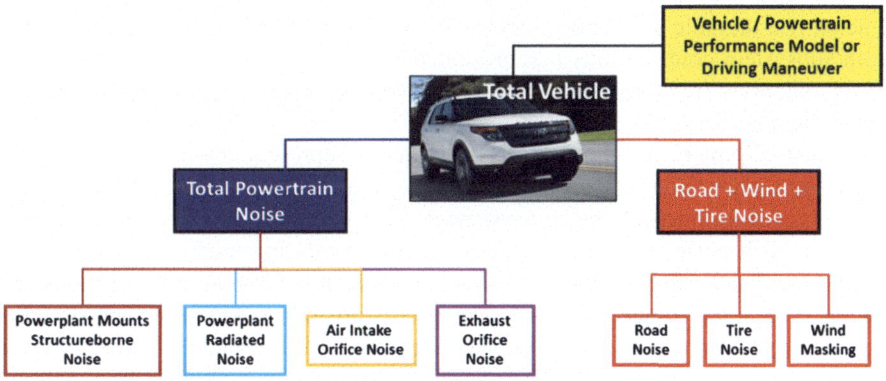

Fig. 3. Hierarchy of DTS model using SPC approach

2.2 Desktop Simulator Sound Objects

The collection of data used to represent the sound of a particular contribution has been termed a "sound object" by B&K. Sounds objects may include harmonic data, e.g. a set of engine orders, masking data, e.g. road/wind noise waveform or spectra, or event-based data, e.g. engine start-up sound time history [2]. Powertrain subsystem contributions are modeled using an SPC sound object that allows the engineer to specify source-level data and a corresponding transfer function. An overview of the contribution modeling approach for each of the significant powertrain sources is provided in the following section.

The B&K DTS toolset includes data preparation utilities to decompose measured time signals into harmonic and masking components. Orders (of engine, wheel, or other rotor speed) are extracted based on a tachometer signal and then subtracted from the original signal to yield the random, or masking, sound [3]. For some powertrain airborne sources, both the orders and masking content can be significant contributors. In such cases, both harmonic and masking SPC sound objects are included in the DTS model. Coherent order-based sound objects in a DTS model should be phase-aligned to yield accurate total sound when combined.

For fully interactive (drivable) simulator models, sound objects represent noise contributions for a range of operating conditions: engine speed, engine load, transmission gear, etc. The data are organized as matrices and/or arrays internally within the Desktop Simulator software, and interpolation is performed based on the driving conditions to synthesize the appropriate sound.

2.3 Defining Powertrain Subsystem Noise Contributions

Source data for powertrain noise contributions can be measured at the system or subsystem level via powertrain dynamometer rig testing or through vehicle testing (on-road or in a chassis rolls dynamometer cell). Measurement data can also be adjusted based on CAE analyses to represent new or modified powerpacks and subsystem/component designs. Several engine load conditions should be measured/predicted if a drivable simulator model is to be constructed. The noise path for powertrain subsystem contributions is represented by vehicle acoustic and vibro-acoustic transfer functions (relating powertrain source levels to driver's ear sound) based on measurement and/or CAE.

Exhaust tailpipe orifice noise (TON) is typically represented with a harmonic SPC sound object whose source data are defined as engine orders from CAE (1-D acoustic model) or from microphone measurements near the tailpipe orifice. In some cases, exhaust air rush noise (flow noise) may be a concern during heavy acceleration. The tailpipe broadband noise for the conditions of concern may be included as a separate masking type sound object. TON measurements are conducted in a hemi-anechoic powertrain dynamometer test cell that accommodates a full vehicle exhaust system, or via on-road testing of a vehicle. The interior contribution is computed by combining the TON source data with the vehicle acoustic transfer function (typically measured on the vehicle) representing the noise attenuation from the tailpipe orifice to driver's ear.

For Air Induction Systems (AIS), the Intake Orifice Noise (ION) contribution can be represented using an approach similar to the one used for exhaust TON contribution. The ION source level may be measured in a hemi-anechoic powerplant dynamometer or vehicle chassis rolls test. 1-D acoustic modeling may also be employed to calculate the ION source orders for new designs. Acoustic transfer functions are measured on the vehicle to relate the noise at the intake orifice to that at the driver's ear.

Powerplant radiated noise is measured in a hemi-anechoic powertrain dynamometer test cell using microphones placed approximately 1-m from each side of the powerplant. The powerplant is installed in such a manner to minimize AIS and exhaust system noise contributions (orifice noise and shell radiated noise) to the microphone measurements. Acoustic transfer functions representing noise attenuation from the 1-m microphone locations to driver's ear are measured on the vehicle. Both harmonic and masking sound objects are used to represent the powerplant radiated noise contribution from each side of the powerplant.

Powerplant mount structureborne noise contributions are represented using harmonic sound SPC sound objects. Source data are defined as mount forces (in 3 directions at each mount attachment point to the body) for the orders of interest. Powerplant mount operational forces for desktop simulator models are typically estimated using the stiffness method (Hooke's law). Powerplant mount active-side vibration may be measured in a powertrain dynamometer test setup (and measured concurrently with powerplant radiated noise data), in a vehicle during on-road or chassis rolls testing, or it may be predicted using CAE analysis. Powerplant mount stiffness is defined as a function of quasi-static preload and frequency using Ford proprietary tools and targeted (or measured/predicted) mount characteristics. Body or frame structural compliance at the mount attachment points can also be incorporated into the estimation of the total stiffness. Mount forces are computed by combining the load- and frequency-dependent mount stiffness tables for each operating condition of interest (engine load, gear, etc.) with the corresponding active-side mount displacement order data. The SPC sound object combines calculated mount forces with vibro-acoustic transfer functions representing driver's ear sound in response to force applied at the mount-to-body/frame interface points. The transfer function data may be acquired from vehicle testing, CAE analysis, or a combination of the two.

2.4 Road and Wind Noise Modeling

Surrogate road, tire, and wind noise measurements are typically used in simulator models for powertrain NVH assessment early in the product development process. Total road, tire, and wind noise can be measured in a vehicle during a neutral coast-down condition. Coherence-based methods can be used to decompose measured interior noise into various contributions from road, tire, and wind noise if required [2]. The data are assigned to masking sound objects in DTS to represent noise for a range of vehicle speeds. Tire and road noise data for multiple road surfaces may also be incorporated into the model.

2.5 Vehicle Performance Modeling

A vehicle performance model or fixed driving maneuver must be defined to govern playback of the sound objects and the visual elements such as the road and surroundings, tachometer and speedometer displays. For a free-driving simulation, the vehicle performance model also controls how the simulator model responds to driver inputs including accelerator/brake pedal and manual upshift/downshift requests. For a fixed driving maneuver a set of time traces that describes engine speed and load, vehicle speed, transmission gear, etc. is defined for the event of interest, e.g. part throttle acceleration at 0.1 g in a fixed gear. This information could come from a vehicle test, a simple physics-based performance model, or from a complex multi-domain simulation with detailed models of the powertrain and vehicle, including relevant control systems.

Setting up a model for evaluation with a fixed driving maneuver has several advantages compared to a free-driving model. First, since sound objects need only be defined for the conditions corresponding to the fixed driving maneuver, the time required to build the model is reduced relative to a free-driving model. Additionally, depending on the program maturity, it is sometimes the case that powertrain source-level predictions/targets will only be available for certain conditions, e.g. full load. Another advantage is that the auditory and visual stimuli during an evaluation are completely repeatable, eliminating the possibility of evaluators reaching different conclusions due to inconsistent driving behavior. A disadvantage of this approach is that it does not provide information about how driving style/behavior could influence the rate of occurrence and perception of a particular powertrain noise.

A free-driving model provides a more immersive experience; however, SPC sound objects in a free-driving model must represent powertrain source-level data for the full range of operating conditions, e.g. engine speed and load, and can require significantly more time to prepare compared to those found in a model with a single driving maneuver. The performance model within DTS requires one to define vehicle speed vs. time traces for constant pedal (or engine load) run-up and coastdown conditions in each transmission gear. Depending on the intended purpose of the model/evaluation, this information may come from a Ford proprietary vehicle performance modeling toolset, or a simple physics-based performance model such as the one that is included with the B&K DTS toolset. The DTS software continuously interpolates between these predefined curves based on driver inputs during a simulation to determine the powertrain/vehicle operating conditions.

A drawback of using the DTS performance model for free-driving is that it may not always be able to represent powertrain/vehicle behavior for many modern applications with complex powertrain hardware and control system logic. Examples include torque converter function in conventional automatic transmissions (fluid coupling only vs. slipping converter clutch vs. locked converter clutch), complex gearshift scheduling logic (as found in newer Ford 8- and 10-speed transmission applications), as well as the operation of Continuously Variable Transmission (CVT) and hybrid electric vehicle (HEV) powertrains. For some evaluators, unrepresentative behavior of the performance model can distract from the primary objective of the NVH study.

A more accurate free-driving simulation has been achieved by linking DTS with Ford proprietary multi-domain Vehicle Model Architecture Powertrain Simulation (VMAPS) models. The DTS and VMAPS models run simultaneously on the same PC and communicate using User Datagram Protocol (UDP) packets. The Ford VMAPS models are typically unavailable until later in the product development process, which necessitates using surrogate models for early development exercises.

3 Application Examples of SPC Desktop Simulator Modeling Approach

3.1 Validation of Approach

As the Ford PT NVH simulator strategy was being developed, evaluator feedback from initial studies indicated skepticism about the representativeness and overall balance of powertrain and road/wind noise. In order to validate and build confidence in the approach, two DTS models of a small CUV (in production) with turbocharged 4-cylinder engine were developed. One model was built with powertrain source and transfer function data using SPC sound objects, which is described within Ford as a Level 3 model. The other model was created based on direct interior noise measurements from on-road vehicle testing, which is termed at Ford as a Level 1 model.

The source data for the Level 3 model consisted of powertrain dynamometer rig measurements for powerplant radiated noise and mount vibration, as well as vehicle measurements for ION and TON source levels. These data were available from standard tests that had been performed previously during program development. Likewise, vehicle airborne and structureborne noise transfer functions had also been measured as part of standard testing during program development. Interior noise measurements for the Level 1 model were conducted on-road using an early production-level vehicle. Note that different vehicle specimens were used for the ION and TON source measurements, transfer function measurements, as well as the direct interior noise measurements for the Level 1 model. This introduced part-to-part/vehicle build variability into the study.

A comparison of the full-load interior noise spectrograms generated by the Level 3 and Level 1 models is shown in Fig. 4. The major engine order content (2, 4, 6, etc.) visible in the Level 1 model spectrogram is present in the Level 3 model spectrogram and at similar levels. The level of broadband noise is also similar. There are engine orders/frequencies with high energy in the Level 1 model that do not appear in the Level 3 model. This could be due to other noise paths that aren't represented in the Level 3 model (half shaft torsional through suspension, driveline structureborne, exhaust hanger structureborne, heater hoses and fuel lines structureborne, etc.), or due to differences in the source levels and transfer functions between the specimens used for the Level 1 and Level 3 models. Despite these differences in order levels, the interior loudness levels match relatively well between the Level 3 and Level 1 models as shown in Fig. 5.

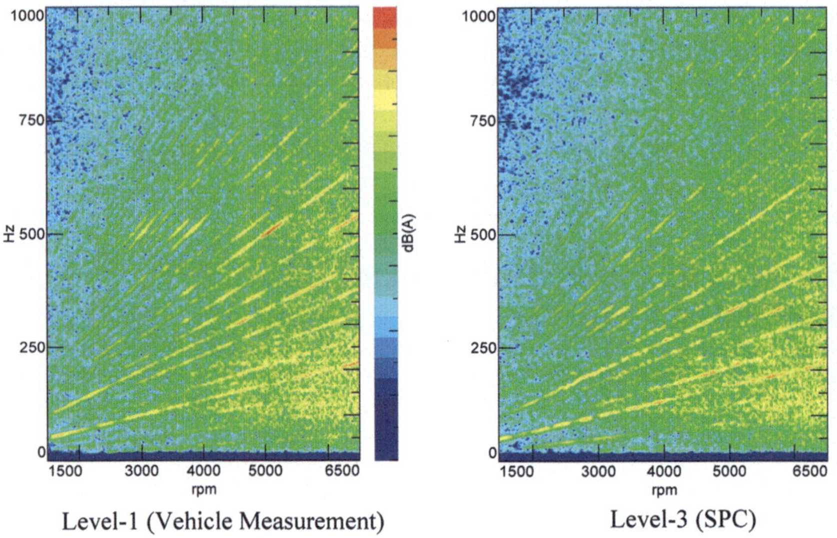

Fig. 4. Interior SPL spectrograms for Wide Open Throttle (WOT) simulation in Level 1 and Level 3 DTS models.

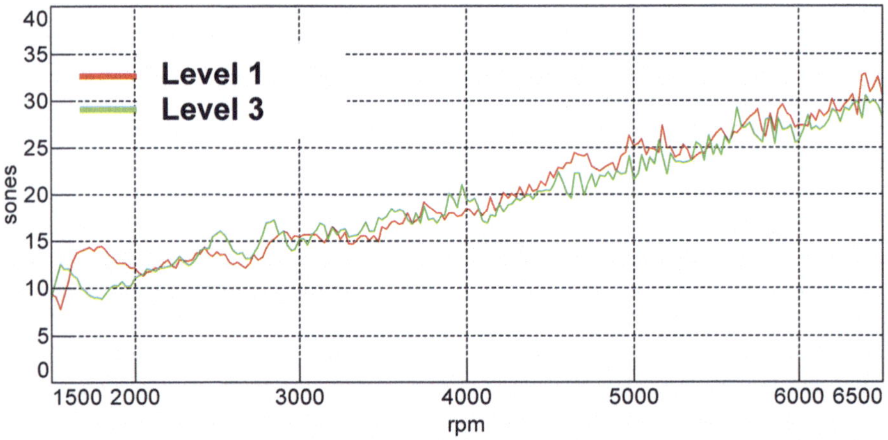

Fig. 5. WOT loudness predictions for Level 1 and Level 3 DTS models.

Listening studies were conducted with powertrain NVH subject matter experts and management to demonstrate the capability of the Level 3 DTS modeling approach. Evaluators listened to the models back-to-back and the majority were unable to distinguish the Level 3 from the Level 1 model. While most could identify some subtle differences, the consensus was that the Level 3 model very closely represented the overall sound character and level of the actual vehicle and could be used for powertrain content and hardware development studies.

3.2 Additional AIS Resonator Study

During the early stages of program development for a new midsize pickup truck, CAE predictions (1D acoustic analysis) of intake orifice noise for a naturally aspirated V8 engine installation indicated that 1.5 and 2.5 engine order levels were projected to exceed targets that had been cascaded from the Vehicle Engineering organization as shown in Fig. 6. CAE analysis indicated that an additional Helmholtz resonator tuned to 140 Hz could reduce ION orders to target level; however, the vehicle program was opposed to the piece price increase. Due to some uncertainty regarding the importance of the 1.5 and 2.5 order targets to overall interior sound quality, it was agreed that the desktop simulator should be used to subjectively assess the impact of the proposed resonator.

A Level 3 desktop simulator model was constructed using powertrain source-level data from measurements of the current product. To evaluate the impact of the proposed AIS resonators, the differences in CAE predicted ION order levels at WOT without and with the resonator were applied to the ION levels measured on the current AIS system. A team of evaluators experienced simulated 3rd gear WOT run-ups representative of the current AIS (with 1.5 and 2.5 order target overages) as well as the AIS with proposed resonator (1.5 and 2.5 order targets met). The evaluators agreed that the interior noise differences between the two systems were not discernable. The predicted interior overall SPL for the ION contribution as well as the total (all contributions) shown in Fig. 7 indicates that the 1.5 and 2.5 order ION target overages have only a small influence on the total overall SPL. The team concluded that the 1.5 and 2.5 order targets could be relaxed, and the additional resonator should not be included in the AIS design thereby avoiding the piece price increase.

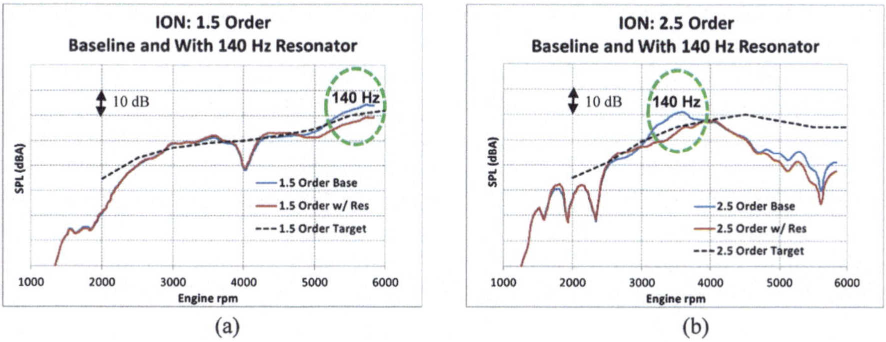

Fig. 6. Predicted WOT ION order levels and targets for a V8 engine application in a new midsize pickup truck: (a) 1.5 engine order, (b) 2.5 engine order.

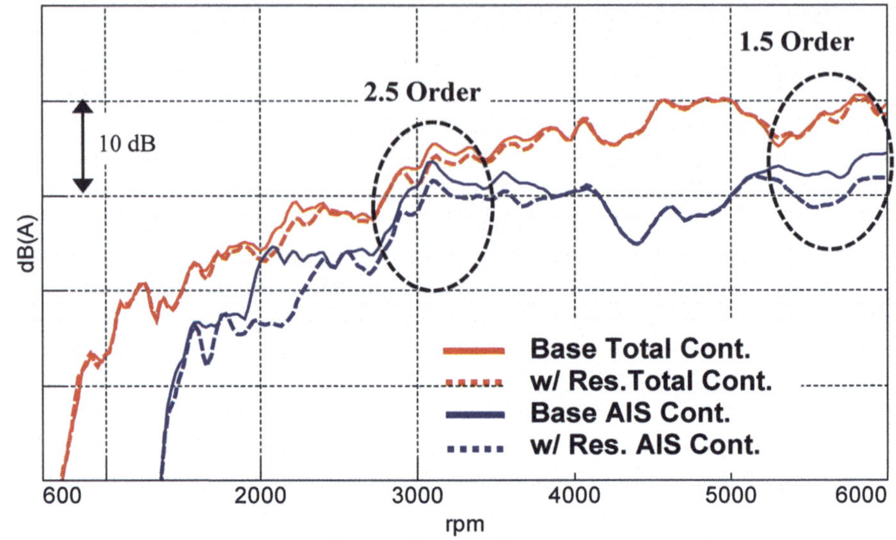

Fig. 7. Predicted interior overall SPL for ION contribution (AIS Cont.) and all contributions (Total Cont.) for AIS designs with (w/Res) and without (Base) resonator tuned to 140 Hz.

3.3 Exhaust System Selection Study

In order to achieve targeted power for a V6 powertrain installation in a new mid-size pickup truck program, it was required to reduce exhaust system backpressure relative to the current design. To reduce backpressure larger exhaust pipe diameters are required, which generally increases low-frequency TON engine order levels due to decreased expansion ratios between pipes (relative to muffler tuning volume). To maintain adequate control of the TON orders, several new exhaust system configuration alternatives with redesign/relocated mufflers were developed. The maximum volume of the muffler shells was limited by underbody package constraints. Prototypes for several of the systems were fabricated and tested in a powertrain NVH dynamometer cell with the current production engine, which provided acoustic excitation and flow very similar to the new engine. To evaluate the NVH performance of these systems in the full-vehicle environment and align on a design for the first prototype vehicle build phase, the measured tailpipe noise orders were incorporated into a Level 3 desktop simulator model using harmonic SPC sound objects.

A team of evaluators evaluated each of the exhaust systems in back-to-back simulated part load (0.1 g vehicle acceleration) and full load run-ups in 3rd gear. It was agreed that two of the configurations (A and D in Fig. 8a) demonstrated an objectionable level of boom during the run-ups. Configuration C was found to be acceptable and exhibited the lowest level of boom; however, its larger muffler volume created a package conflict. Configuration B was also deemed acceptable, and since its smaller muffler volume represented a cost and weight reduction relative to configuration C and did not create a package conflict, it was the design selected for the first prototype build.

Due to significant package constraints imposed by the presence of a high-voltage battery pack, configuration D was the only exhaust system compatible with the HEV variant. Since this configuration was initially deemed unacceptable for boom during the simulator listening study, additional 1-D acoustic analysis and optimization was conducted to improve the order attenuation capability. Figure 8b shows the improvements in predicted interior noise overall level with the optimized systems (E and F), and performance for both was deemed acceptable in subjective simulator evaluations.

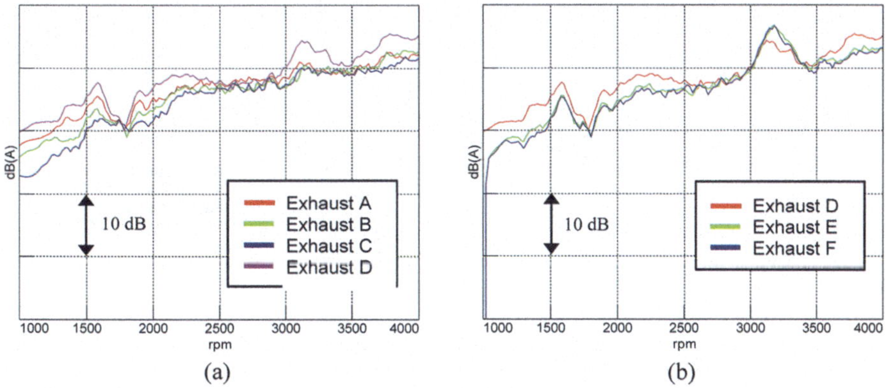

Fig. 8. Predicted overall interior SPL at full load for a V6 engine application in a midsize pickup truck with **(a)** various prototype exhaust system proposals measured in a powertrain dynamometer rig, and **(b)** with CAE optimization of one of the exhaust designs.

3.4 Study of Balance Shaft Removal from Inline 4-Cylinder Engine Application

I4 engines have an inherent 2nd order vertical imbalance force that can contribute to interior loudness and tactile coarseness at high engine speeds. Twin counter-rotating balance shafts are commonly incorporated into the design of larger I4 engines to internally balance the shaking force; however, these systems add cost and weight and increase engine friction losses. For these reasons, the team developing a new large I4 engine for a small crossover utility HEV requested the Powertrain NVH organization to investigate the implications of deleting the balance shafts.

Mount active-side vibration levels for a full load engine speed sweep were predicted for the HEV powerplant both with and without the balance shafts using forced frequency response analysis of a flexible powerplant model. The differences in the CAE predictions for 2nd engine order active-side mount vibration were then used to adjust measurement data from a surrogate legacy powerplant with balance shafts. A comparison of the mount vibration levels can be found in Fig. 9a. The adjusted and unadjusted mount vibration order levels were used to create SPC sound objects in a desktop simulator model, and a team of subject matter experts evaluated the configurations in back-to-back run-ups representative of a Wide Open Pedal (WOP) maneuver for the HEV with electronic CVT type transmission. The evaluators noted significantly increased sound level at 5400 rpm and agreed that it was unacceptable. However, the increase in sound level at all other engine speeds was deemed acceptable. Figure 9b shows the predicted interior overall sound levels with and without balance shafts.

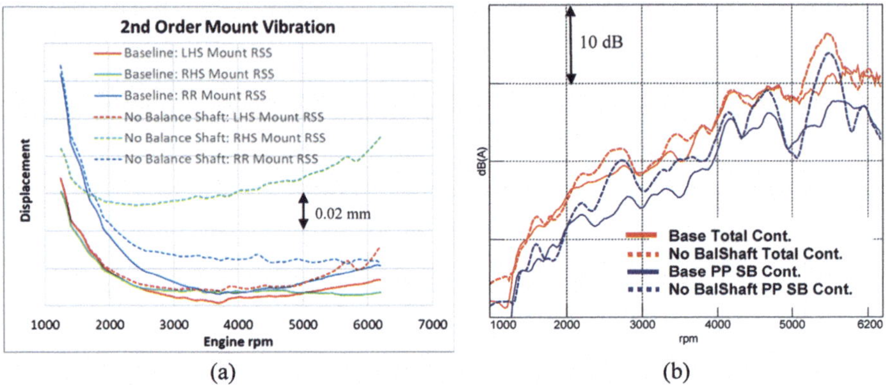

Fig. 9. (a). Active-side powerplant mount vibration predictions for a 2.5 L naturally aspirated I4 engine with and without 2nd order balance shafts. **(b).** Interior overall SPL predictions (powerplant mount structureborne contribution and all contributions) with and without the engine 2nd order balance shafts.

The recommendation to the engine development team was to retain the balance shaft assembly; however, it was also suggested that an opportunity to remove the assembly might exist if the engine could be calibrated to run only as high as 5200 rpm. Due to negative impact on vehicle acceleration capability during a WOP maneuver this was not studied in detail.

3.5 Electronic Drive Unit Motor Whine Improvement

CAE analysis of the initial design for a primary drive unit for a new battery electric vehicle (BEV) indicated that active-side mount vibration levels were significantly exceeding generic targets for motor 24th order (electromagnetic torque ripple). A number of design actions were proposed to reduce the vibration levels, including: heavy ribbing on the motor housing casting, stator reinforcement, strengthening of the power electronics module mounting brackets, and a tuned vibration absorber. The CAE model was leveraged to predict mount vibration levels for three combinations of these design actions. A DTS model was constructed to evaluate interior structureborne gear whine levels in the presence of masking noise for the proposed design actions.

Surrogate road, tire, and wind masking noise was taken from an early production version of a new vehicle with a similar footprint and body profile. The structureborne motor whine contribution was calculated based on the CAE predicted mount vibration, targeted dynamic stiffness values for the electronic drive unit mounting system, and a surrogate noise transfer function based on measurements of a production vehicle with similar rear subframe architecture.

Figure 10 shows interior noise levels predicted by the DTS model for 24th order motor whine for a light acceleration driving maneuver as well as a WOP driving maneuver. The stator reinforcement significantly reduced interior whine levels for a substantial portion of the motor/vehicle speed range of concern for both maneuvers. Additional improvement was observed, primarily for the WOP condition, when the improved mounting brackets for power electronics unit and tuned vibration absorber

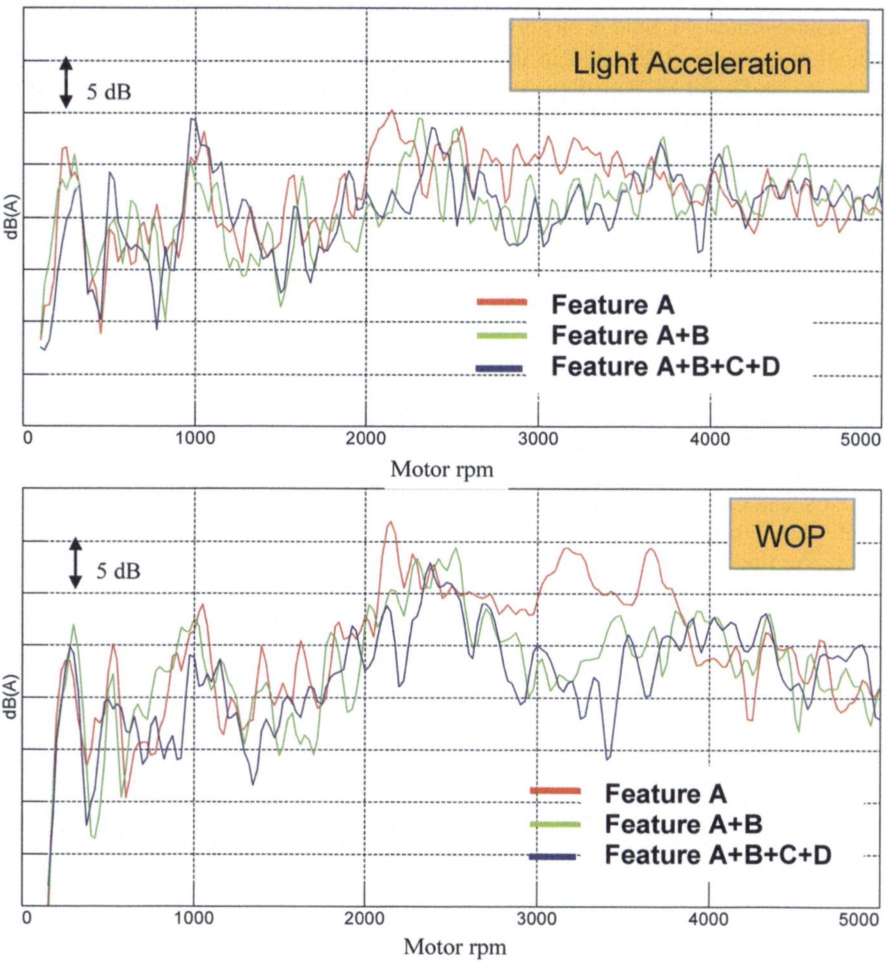

Fig. 10. Electric motor 24th order structureborne contribution to interior SPL for various drive unit design proposals. Feature A: heavy ribbing on motor housing casting; Feature B: stator reinforcement; Feature C: structural reinforcement of the power electronics unit mounting brackets; Feature D: tuned vibration absorber.

were added. The team of evaluators agreed to pursue the heavy ribbing and stator reinforcement design configuration and reconsider the other actions if needed based on vehicle evaluations later in program development.

4 Conclusion

Desktop vehicle NVH simulator models constructed using the SPC approach enable powertrain NVH evaluation in the context of a virtual vehicle environment early in the development process before prototype vehicles are available. Interior sound levels and sound character predicted by simulator models using the SPC approach have been demonstrated to match closely with direct interior noise measurements. The application examples presented in this paper highlight just a few of the studies that have been conducted by Ford Powertrain NVH over the past several years to facilitate hardware content and target decisions, which often result in cost and/or weight avoidance or help to align the program team on content that is necessary to satisfy customer requirements. Simulator usage is expected to grow as powertrain NVH design verification testing moves from prototype vehicles to powertrain dynamometer rigs, and an increasing number of assessments are made analytically.

References

1. Allman-Ward, M., et al., The Interactive NVH Simulator as a Practical Engineering Tool, Proceedings of SAE Noise and Vibration Conference (2003).
2. Crewe, A., Distler, H., Heinz, T.: Simulator Sound Objects – A Proposal for an Open Standard on Sound Components for Driving Simulators. Proceedings of SAE Noise and Vibration Conference (2003).
3. Crewe, A., Bogema, D, et al: Sound Decomposition – A Key to Improved Sound Simulation. Proceedings of SAE Noise and Vibration Conference (2003).
4. Williams, R. et al: Using an Interactive NVH Simulator for Target Setting and Concept Evaluation in a New Vehicle Programme. Proceedings of SAE Noise and Vibration Conference (2005).

Parametric Analysis of an Automotive Wheel Arch Acoustic Treatment

Romain Baudson[1(✉)], Thibault Lafont[2], Viswesh Sujjur Balaramraja[1], Francesca Ronzio[2], and Benoit Van den Nieuwenhof[1]

[1] Free Field Technologies, MSC Software Belgium, rue Emile Francqui 9, 1435 Mont-Saint-Guibert, Belgium
romain.baudson@fft.be
[2] Autoneum Management AG, Schlosstalstrasse 43, 8406 Winterthur, Switzerland

Abstract. Recent evolutions in terms of regulations for pass-by noise force Automotive OEM's to build more silent cars than before. To be able to decrease the emitted noise level while meeting other constraints such as weight, space or sound quality, the OEM's and their suppliers have to come up with innovative solutions to reduce the different noise sources and control the noise propagation around the vehicle. In particular, powertrain and tire noise sources are dominant contributors to the exterior sound emitted by vehicles and efficient methods to understand, analyze and reduce their contribution need to be developed. Exterior acoustic simulations are required to find optimum acoustic treatment location and physical properties while offering the possibility to improve and predict final pass-by noise performance before the first complete vehicle prototype is built.

This paper presents a framework for the efficient simulation, including parametric analysis, of wheel arch acoustic treatments. A finite element scheme is used to solve the acoustic propagation of different sources around a car tire while a locally-reacting admittance boundary condition is used to model the wheel arch acoustic treatment.

The parametric analysis focuses on the investigations of acoustic treatment materials, location and geometries, on the tire contribution to the total pass-by noise level. A model reduction strategy based on a condensation of the car surface and exterior acoustic problem on the tire and wheel arch surface is presented as a mean to a drastic speed-up of the computational process, therefore opening the door to a parametric analysis process.

Comparisons between the different designs as well as the computational performance of the proposed approach are presented. Numerical simulations are conducted with the Actran commercial acoustic simulation framework.

Keywords: Pass-by noise · FEM simulation · Reduction strategy · Condensation

W. Siebenpfeiffer (Hrsg.): *Automotive Acoustics Conference 2019*, Proceedings, S. 185–199, 2020. https://doi.org/10.1007/978-3-658-27669-0_13

1 Introduction

The recent changes in the legislation about the exterior noise generated by transportation vehicles emphasis the importance of tire noise in the total noise generated in typical urban driving conditions [1]. OEMS, tire manufacturers and sound package suppliers do have to tackle with a new major or even dominant noise source contributor to the pass-by (PB) noise. This is a new challenge for the sound package supplier that are requested to develop innovative passive acoustic treatments solutions answering the new market needs.

Measurements is a proven methodology to assess the performances of new design configurations. But the associated time and costs limit the number of comparisons that can be performed in a given timeframe. During the past decades, numerical modelling methods have emerged as powerful tools for the development and the optimization of safer, more efficient and more comfortable vehicles. In particular Finite Element Methods which remain the numerical references, offer to engineers the capacity to evaluate the impact of passive treatments positions together with material performances assessment. Improvements brought to the meshing algorithms and the numerical methods on one hand the increase in computational resources and, on the other hand, allow engineers to solve complex acoustic radiation problems for large systems and relatively high frequencies [2, 3].

This paper presents a new methodology to accelerate the parametric studies of such passive acoustic treatments relying on a static condensation approach. First, the variable part is limited to a subset of the model boundaries, the wheelhouse outer liners. It has the advantage to contain far less degrees of freedom (dofs) than the nominal model, which can be exploited to reduce tremendously the CPU cost of re-evaluating variants of the liner. Then, the model reduction strategy is based on a condensation of the near-field acoustic and exterior acoustic problem to the surface of the acoustic treatments, the outer wheel liners. While this reduction does not bring any benefit when a single configuration is studied, it allows nonetheless a significant reduction of the solution time when a large design space is investigated, which is precisely the case where several outer liner materials are studied.

This article is divided in three sections. First, the numerical simulation set up is described and the material modeling description follows. Then comes the presentation of the static condensation formulation. Here one focused on the process description and the possible extensions. Finally, the third and last section presents the analysis results. Advantages of the static condensation approach in the frame of parametric analyses are presented in terms of computation time and memory consumption. The impact of absorbing materials on the PB overall sound pressure level (OASPL) and related acoustics transfer functions are commented.

2 Numerical Simulation

2.1 Finite Element Model Description

A Finite Element (FE) model of the DrivAer vehicle is built to study the acoustic propagation around the complete vehicle using the Actran Acoustic commercial software [4]. The DrivAer is a generic vehicle model, bare from any geometrical details. It includes all the important features of typical car architecture such as engine bay and underbody geometry used on both academic and industrial domains for advanced FE simulations investigations [3, 5, 6].

Pass-by noise results can be obtained using a frequency domain simulation by modeling a fixed vehicle located in the middle of the test zone [7]. Two lines of 30 microphones each placed 30 cm from each other are defined on each side of the vehicle respectively at 7.5 m from the track center and at 1.2 m from the ground. Figure 1 illustrates the simulation setup.

Fig. 1. DrivAer model in numerical pass-by noise calculation setup

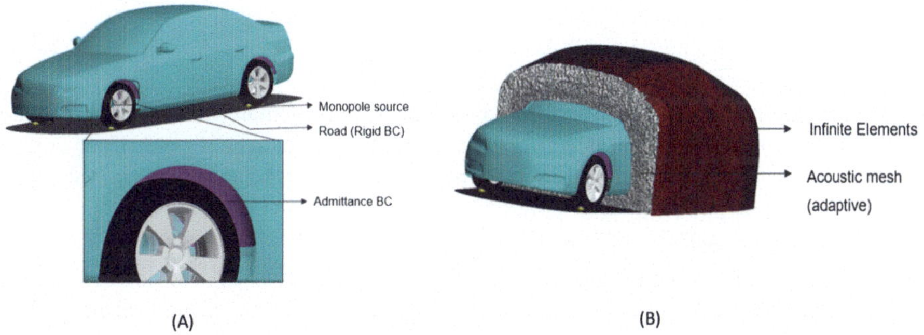

Fig. 2. (A) DrivAer model with boundary conditions (B) Automatically generated acoustic and infinite elements.

The near-field acoustic field is modelled using a layer of fluid around the car. Far-field acoustic field is studied by using Infinite element (IFE) method which helps to model unbounded acoustic domain (cf Fig. 2). The acoustic mesh is using a h-adaptivity feature available within the Actran software. The solver is then able to automatically generate meshes for the different frequency ranges ensuring an optimized number of dofs to be solved.

The road is modeled as a perfect rigid plane, where the velocity is zero. The sources used are monopoles which are located at the vicinity of the tire/road contact patch. Such sources are easy to handle in an FE simulation and, as long as they remain sufficiently close to the contact patch, happen to be represent correctly the radiation of a rolling tyre notably at the frequency range of interest for PB noise, above 800 Hz [8].

2.2 Acoustic Wheel Treatments

The presence of the acoustic treatments is modeled by an admittance boundary condition at a specific location in the wheel outer liner. In this case, the upper part of the

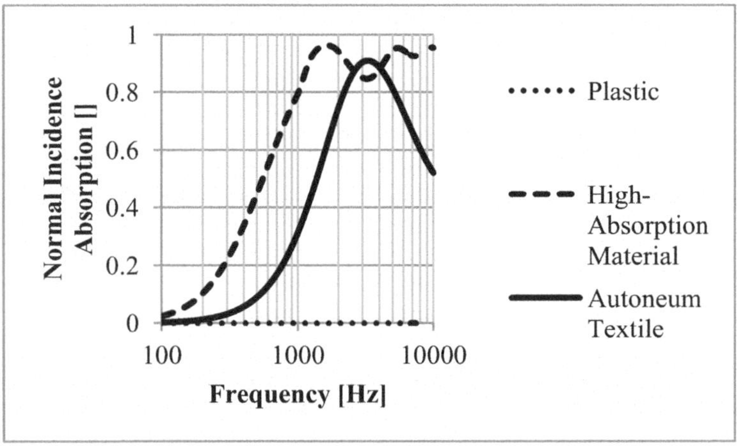

Fig. 3. Normal incidence absorption levels for the different materials used for the wheel outer liner upper region.

wheel arch has been chosen, like shown in Fig. 2. Three materials have been taken for the simulation: a plastic material, a typical treatment for wheel outer liner coming from Autoneum and a high absorptive treatment material. The admittance tables have been directly exported from Autoneum VisualSISAB, a TMM-based [9] software to assess the acoustic performance of multilayers.

The purpose of choosing these three different materials is first to perform a parametrical study by analysis the calculation gain when using the condensation process while moving from one material to another. The second purpose is to see how this upper region of the wheel outer liner is sensitive to an absorption change. Indeed, from Fig. 3 – the normal incidence absorption levels for the different materials used for the wheel outer liner upper region, one sees that at the frequency band [800–1200 Hz] where the tire radiation is the maximum, the three different materials have significant different normal incidence absorption levels. Around 1 kHz, where the radiation of the tyre is supposed to be maximal, the absorption of plastic is negligible, the Autoneum textile is of 0.3 and the high absorptive material is of 0.8.

3 Static Condensation

3.1 Concept

In this paper, a model reduction strategy based on a static condensation approach is studied to assess its performance aspects and accuracy compared to conventional approaches. This condensation approach is expected to reduce tremendously the computational cost during optimization process or parametric analysis [10]. Typically, during parametric or optimization studies, a numerical model is solved for each set of design variables which are the different acoustic treatments. In such cases, finite element models of the near-field acoustic and far-field acoustic problem are not modified, only the admittance boundary is updated for each iteration.

Contrary to conventional methods where one updates the full model every time there's an impedance change, the main idea of static condensation is to construct a

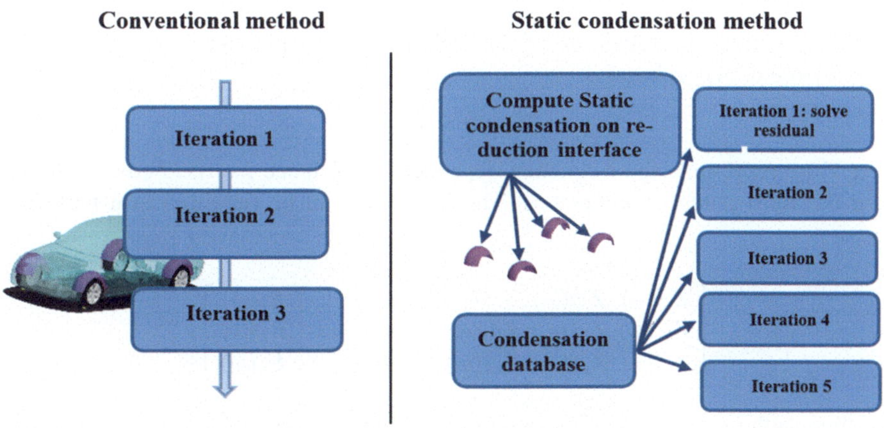

Fig. 4. Conventional method vs. static condensation method – Illustration

Schur complement reduction [11] of the fixed part of the finite element model at the interface with the variable part of the model. In other words, the Schur complement is a matrix reduction operation where the size of the reduced model corresponds to the number of interface dofs. As a result, only a residual problem has to be solved, which contains the updated admittance boundary condition coupled to the Schur complement reduced problem. Figure 4 illustrates this model reduction strategy versus a conventional method.

3.2 Problem Set-Up and Data Recovery

In order to apply a static condensation procedure, the unknowns in the numerical model are classified either as internal unknowns, reduction unknowns or residual unknowns. The first category, the internal unknowns, comprises the set of dofs located in the domains that will be condensed. These unknowns are not recovered during the solution procedure, but they drive the Schur complement construction of the reduced model. The second category, the set of reduction unknowns, represents the fluid pressure or modal participation values at the interfaces between the reduced model and the remainder part of the problem, for example at the interface with the acoustic liners. This set is augmented with a limited number of dofs for recovering the far-field pressure levels in a set of microphones. Finally, the last category, the set of residual unknowns, is not considered in the static condensation procedure. The residual unknowns correspond to the parts of the model that are not reduced and which can be modified independently of the condensed parts. These unknowns intervene for assembling the residual problem and can be coupled to the reduction unknowns. The algebraic operations related to the static condensation computation procedure are detailed hereafter.

3.3 Static Condensation Procedure

The finite element discretization of the Helmholtz equation can be characterized by the following matrix equation:

$$\boldsymbol{Z}(\omega)\boldsymbol{u} = \boldsymbol{F}(\omega), \tag{1}$$

where $\boldsymbol{Z}(\omega)$ represents the frequency-dependent impedance matrix, $\boldsymbol{F}(\omega)$ the loading and $\boldsymbol{u}$ the fluid potential unknowns. This equation can be rewritten in a partitioned way as:

$$\begin{bmatrix} \boldsymbol{Z}_{ii}(\omega) & \boldsymbol{Z}_{ib}(\omega) & 0 \\ \boldsymbol{Z}_{bi}(\omega) & \boldsymbol{Z}_{bb}(\omega) & \boldsymbol{Z}_{be}(\omega) \\ 0 & \boldsymbol{Z}_{eb}(\omega) & \boldsymbol{Z}_{ee}(\omega) \end{bmatrix} \begin{bmatrix} \boldsymbol{u}_i \\ \boldsymbol{u}_b \\ \boldsymbol{u}_e \end{bmatrix} = \begin{bmatrix} \boldsymbol{F}_i(\omega) \\ \boldsymbol{F}_b(\omega) \\ \boldsymbol{F}_e(\omega) \end{bmatrix}, \tag{2}$$

according to the set of internal (i), reduction (b) and residual (e) unknowns. The static condensation approach consists of eliminating the internal unknowns from this system through a Schur complement construction. This results in the residual problem for the reduction and residual unknowns:

$$\begin{bmatrix} \boldsymbol{S}(\omega) & \boldsymbol{Z}_{be}(\omega) \\ \boldsymbol{Z}_{eb}(\omega) & \boldsymbol{Z}_{ee}(\omega) \end{bmatrix} \begin{bmatrix} \boldsymbol{u}_b \\ \boldsymbol{u}_e \end{bmatrix} = \begin{bmatrix} \boldsymbol{R}(\omega) \\ \boldsymbol{F}_e(\omega) \end{bmatrix}. \tag{3}$$

The reduced impedance matrix $\boldsymbol{S}(\omega)$ and reduced loading $\boldsymbol{R}(\omega)$ are respectively defined as:

$$S(\omega) = Z_{bb}(\omega) - Z_{bi}(\omega)(Z_{ii}(\omega))^{-1}Z_{ib}(\omega), \tag{4}$$

$$R(\omega) = F_b(\omega) - Z_{bi}(\omega)(Z_{ii}(\omega))^{-1}F_i(\omega). \tag{5}$$

The construction of the reduced matrices is independent of the equations in the residual unknowns. When solving the reduced residual problem for different conditions on the residual dofs, the same solution is retrieved as when solving the equivalent full finite element system. The proposed model reduction strategy does not require to apply any approximation to the original system.

In order to set-up the reduced impedance and loading matrices, a matrix inversion with the impedance matrix on the internal dofs is required. This is computed efficiently by using a sparse direct solver such as MUMPS [4] which enables a direct computation of the Schur complement matrix of the condensed right-hand side.

The reduced matrices are dense matrices. In the case of a large number of reduction dofs, the dense reduced matrix can also be computed in an out-of-core procedure. After factorizing the sparse impedance matrix $Z_{ii}(\omega)$ once, the Schur complement is computed by a block of columns, where for each block, only a back-transformation is required. In this way, the entire reduced matrix does not need to be stored in memory during the static condensation step.

The assembled residual problem contains sparse impedance matrices on the residual unknowns coupled to a dense static condensation matrix. When the number of residual dofs is relatively low compared to the number of reduction dofs, solving the assembled residual problem can be done most efficiently by a dense direct solver such as LAPACK.

3.4 Extensions to the Static Condensation Approach

The static condensation approach requires to construct a new reduced model for every computational frequency. When the reduced impedance and loading matrices are computed for a predefined set of reduction frequencies $\omega_1, \omega_2, \ldots, \omega_n$, but the solution to the residual problem is required at a different frequency ω_s, the reduced model can be approximately constructed using linear interpolation:

$$S(\omega_s) = S(\omega_1) + \frac{\omega - \omega_1}{\omega_2 - \omega_1}(S(\omega_2) - S(\omega_1)), \tag{6}$$

where ω_1 and ω_2 are the closest reduction frequencies to the solution frequency. The solution to the residual problem corresponds approximately to the solution of the original model. The quality of the approximation depends on the selected reduction frequencies and on the physics of the domains that are included in the reduction.

Table 1. Computer specifications

Processor	32x Intel(R) Xeon(R) CPU X7560 @ 2.27 GHz
Physical memory/RAM	254 GB

4 Analysis

4.1 Performance Aspects

The performance aspects of a static condensation method are studied by comparing computational time and memory consumption with respect to a conventional method. All results presented in this paper have been obtained on a computer system whose specifications are presented in Table 1.

For both conventional and static condensation method, the meshes using h-adpativity are created by the solver in a preliminary step, the mesh creation.

Figure 5 compares the computational time for different execution sequence for conventional method and static condensation method. One can observed that during the initial run of the analysis, the total time consumed by the static condensation method is slightly higher than the conventional method. This is predominantly because of the model reduction (or the static condensation) sequence. This sequence is carried out only once for a particular model, and during the following iterations of the analysis, only the solution sequence is executed. It is observed that the time consumed for solution sequence with static condensation method is drastically reduced, which is highly advantageous while re-running the model with new modified boundary condition. The computational time gain obtained during the re-run of the model with static condensation method compared to conventional method is illustrated in the Fig. 6. In this study, average computational time per frequency in each frequency range of computation is used for comparison.

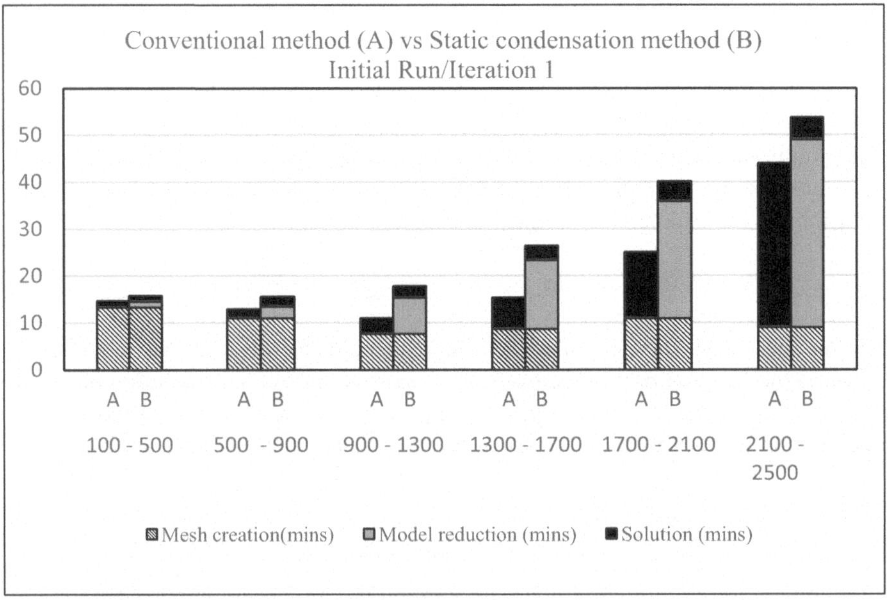

Fig. 5. Comparison of computational time for different execution sequence for conventional method (A) and static condensation method (B)

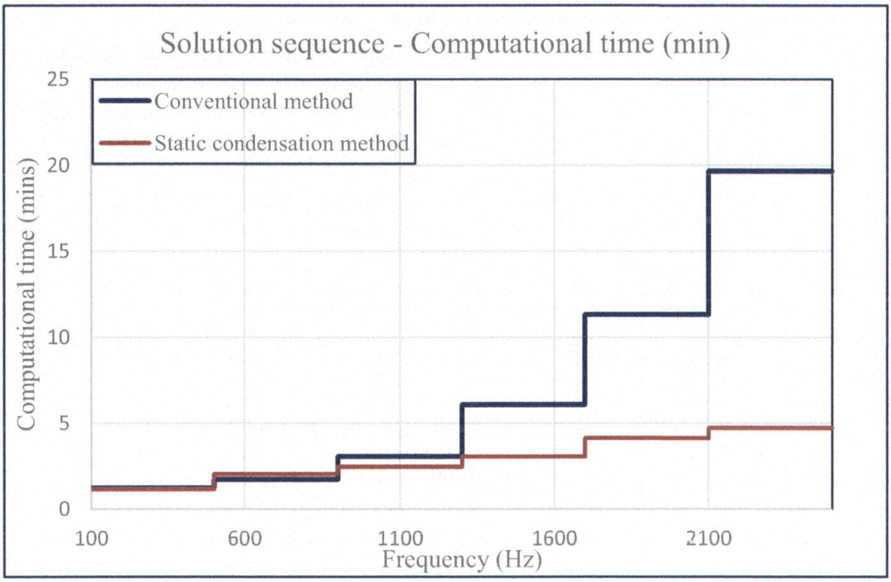

Fig. 6. Comparison of computational time for different solution sequence during rerun of model with different impedance BC

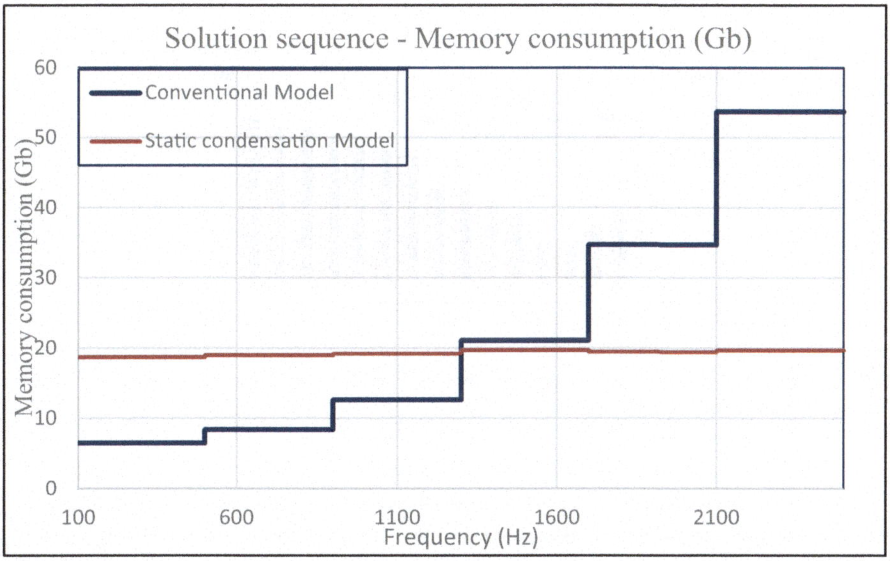

Fig. 7. Comparison of computational memory consumption between two approaches for solution sequence

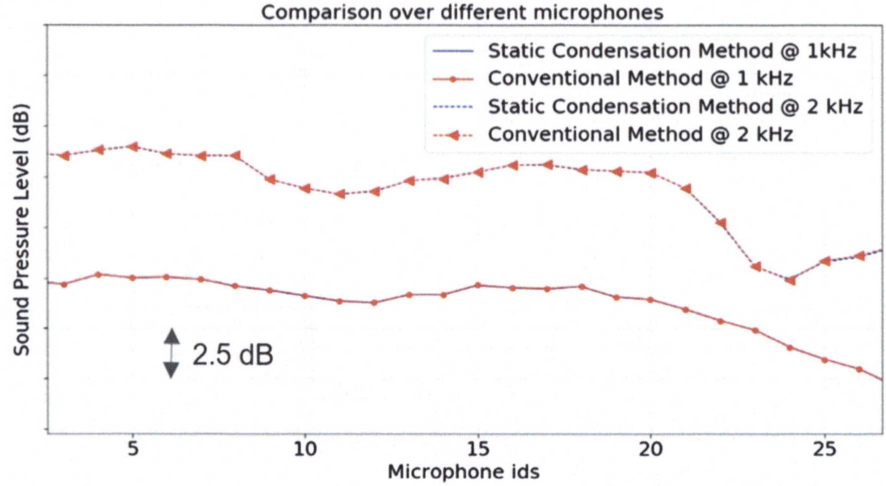

Fig. 8. Comparison of static condensation approach with conventional method at different microphones

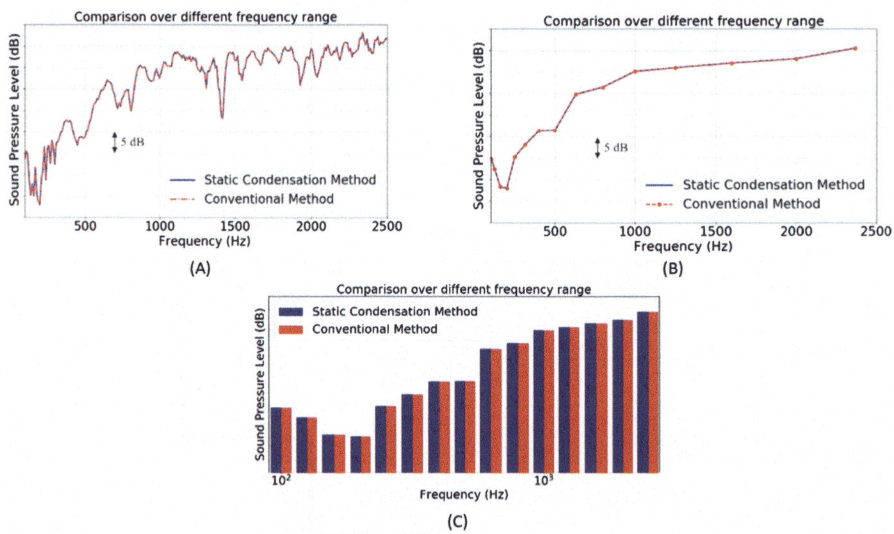

Fig. 9 Comparison of static condensation approach with conventional method over different frequency range at microphone 1

The static condensation method is almost 60% more time efficient than the conventional method. In addition, the memory cost for the static condensation method is almost uniform over the all frequency range. On the opposite, the memory consumption of conventional approach increases with the frequency range as the dofs to be solved increases proportionally. This makes the static condensation approach advantageous on computational resources at higher frequency range, see Fig. 7.

4.2 Accuracy

The static condensation method has its advantage over the conventional method in performance aspects as seen in the previous section. It is necessary to assess the accuracy of the static condensation method. For this aspect, the Sound Pressure Level (SPL) obtained over different microphones due to unitary monopole source excitation at the front of front-left tire has been analyzed and results reported in Figs. 8, 9 and 10.

Since, the primary focus is to compare the two methods, the absolute levels are not the focus of interest. It can be observed that the static condensation method is as accurate as the conventional approach confirming the high benefit of the approach for this analysis type.

4.3 Parametric Analysis

The parametric studies are carried out by modifying the impedance boundary condition using the three different materials defined in Sect. 2. Effect of different boundary conditions are studied using the static condensation database as seen in the Fig. 4. In particular, the different materials performances and effects are compared with respect to the PB noise response. The overall sound pressure level (OASPL) and specific acoustics transfer functions are compared. The nominal configuration is

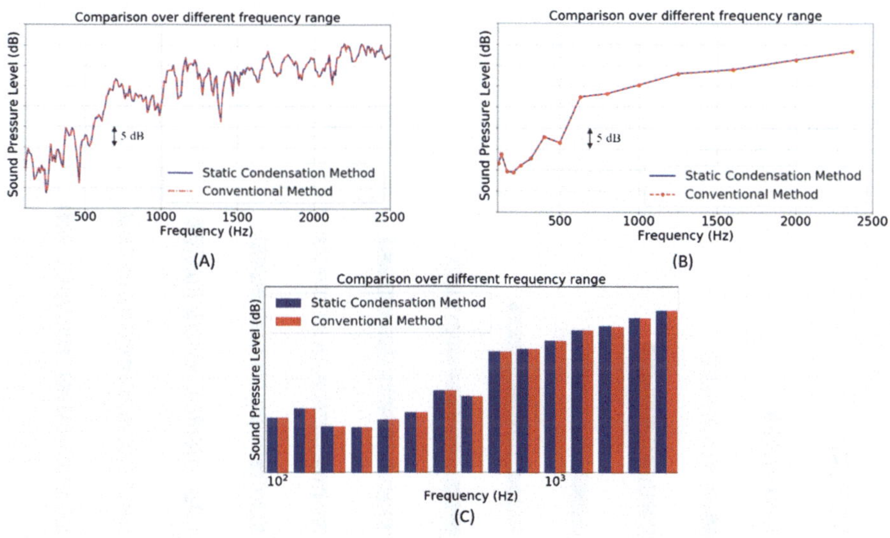

Fig. 10. Comparison of static condensation approach with conventional method over different frequency range at microphone 30

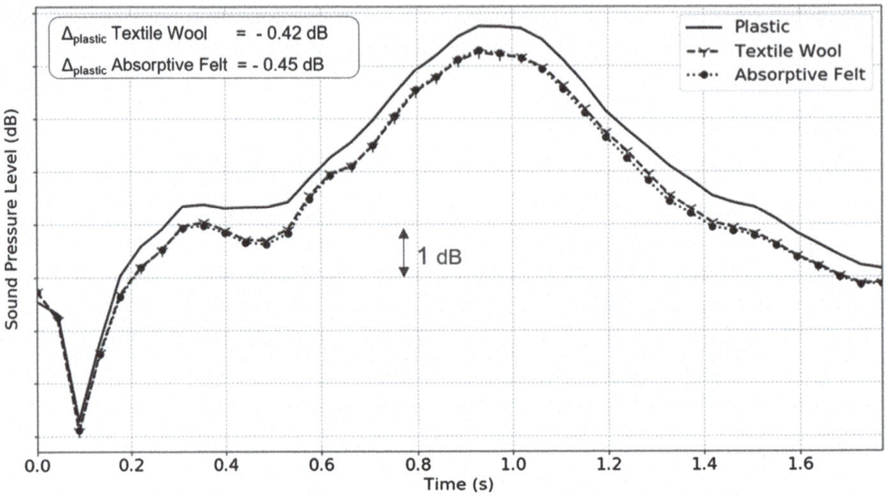

Fig. 11. OSPL – Materials effects comparison

using the plastic admittance properties and is expected to provide the less interesting configuration with respect to a reduction of the OASPL.

Figure 11 compares the OSPL obtained for the three admittance values. One can observe that applying a treatment has an effect on the OASPL. Indeed, the high acoustic material and the Autoneum wheel outer liner material show a similar benefit compared to the plastic material. All in all, the gain is around 0.4 dB which is not negligible given the drastic constrains that require the new PB noise regulation. If one

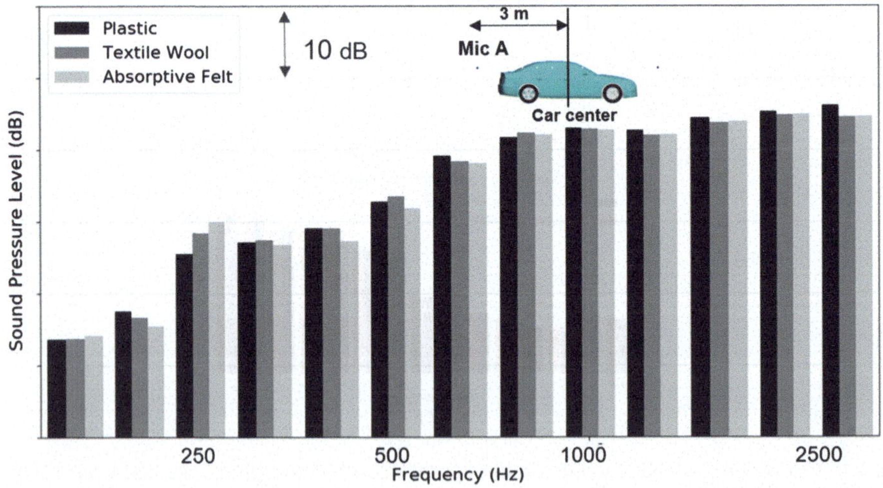

Fig. 12. Acoustics transfer function comparison – Rear microphone, −3 m away from car centered microphone

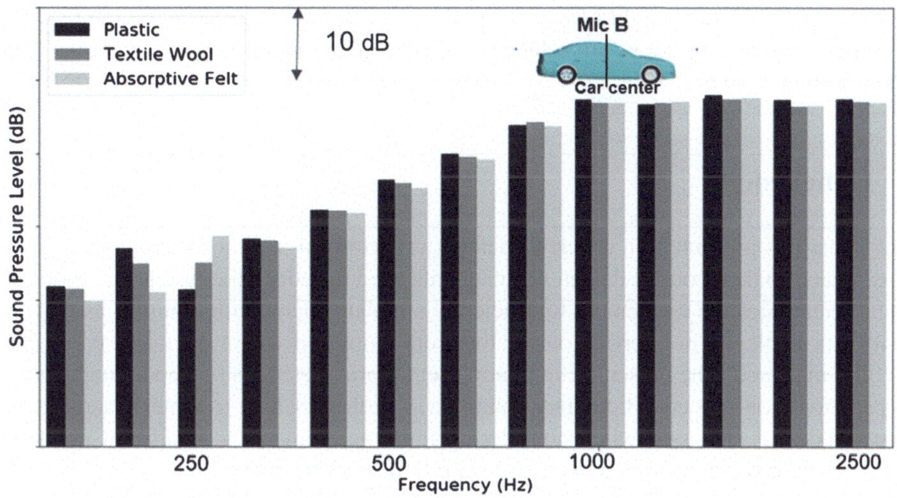

Fig. 13. Acoustics transfer function comparison – Car centered microphone

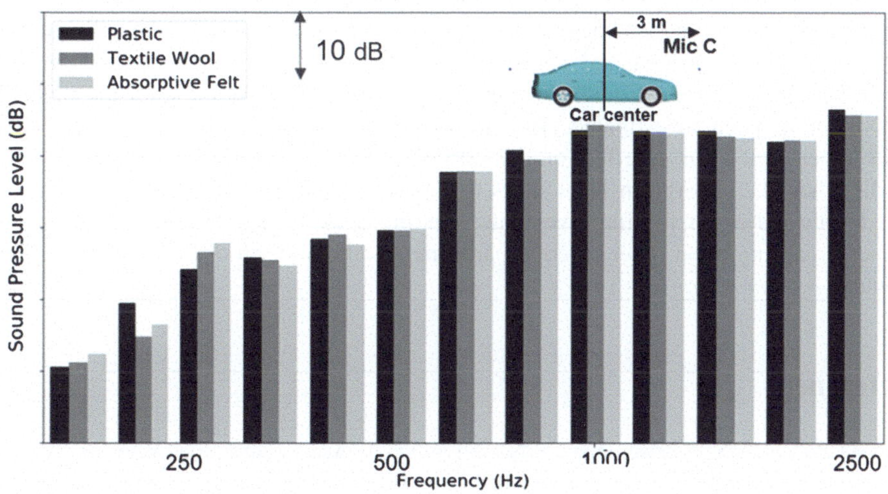

Fig. 14. Acoustics transfer function comparison – Front microphone, +3 m away from car centered microphone

sees an effect from changing from almost a null absorption level (plastic material) to a significant absorption level (Autoneum wheel outer liner material or high absorptive material), this is not the case when moving to a medium absorption level to a high absorption level. A reason for this lack of sensitivity can be the treated region location, which in this paper is the upper one.

To obtain a better insight of the material performances and effects on the OASPL, three PB noise microphones are selected and related acoustics transfer functions compared, Figs. 12, 13 and 14.

5 Conclusion

This paper has presented an efficient method to perform a parametric analysis of problems related to the acoustic design of treatments used to reduce the tyre noise. Finite-element techniques are convenient to efficiently simulate such configurations, as they are easily repeatable, do not depend on weather conditions and where the initial parameters are fully mastered. The presented simulation set up provides tools for modelling acoustic propagation, non-reflective boundary condition, wheel outer liner treatments and pass-by noise modeling. Though, when considering multiple analysis, a significant computational cost is involved as iterations triggers a full model simulation.

A significant benefit of the computational time is obtained by resorting to a static condensation procedure that drastically speeds up the iterative analysis loops. Indeed, this is happening by factoring out of a loop the computational cost related to the part of the model not subjected to design changes.

The Actran simulation environment, featuring this strategy the two improvements presented in this paper, has been used to demonstrate their efficiency. On a 3D generic car model, computational and memory cost reduction have been achieved without loss of accuracy. Three different materials and their related benefits on the final PB OASPL have been investigated. The use of absorbing materials in the wheel outer liner upper region has an effect on the PB noise which is limited to a couple of tenth of dB when the absorption level starts to be significant at 800 Hz and above. There can be always room for improvements: treating also the lower areas of the wheel arches or calling upon other external parts such as the underfloor or the under-engine shields.

References

1. ISO, 362-1:2015, "Measurement of Noise Emitted by Accelerating Road Vehicles – PArt 1: M and N Categories," International Organization for Standardization, Geneva, Switzerland, 2015.
2. R. D'Amico, Stelzer, R., Grebert, J., Chandler, P. and al, "On the Role of Simulation to Account for Design Complexity of Engine Encapsulation," in *Automotive Acoustics Conference (AAC2017)*, Zutich, Switzerland, 2017.
3. T. Lafont, J. Jan Horak, R. D'Amico, R. Stelzer and C. Bertollini, "Passive Treatment Solutions for the Reduction of Vehicle Exterior Tire Noise," *SEA Technical Ppaer,* 2018-01-1571.
4. Free Field Technologies, MSC Software Belgium, Actran 19 User's Guide, Free Field Technologies, MSC Software Belgium, 2019.

5. Hef, A; Indinger, T; Adams, N, "Experimental and Numerical Investigation of the DrivAer Model," in *ASME*, Puerto Rico, 2012, FEDSM2012-72272.
6. Hef, A; Indinger, T; Adams, N, "Introduction of a New Realisitc Generic Car Model for Aerodynamic Investigations," in *SAE Wrold Congress*, Detroit, MI, 2012.
7. A. Talbot and G. Lielens, "Pass-by Noise Synthesis from Frequency Domain Exterior Acoustics Simulations," in *Internatinal ATZ Conference on Vehicle Acoustics*, Zurich, Switzerland, 2017.
8. T. Lafont, R. Stelzer, R. D'amico, W. Kropp and C. Bertollini, "Modelling tyre noise in finite element simulations for pass-by noise prediction," *Journal of Mechanical Engineering Science,* Vols. 203–210, 1989–1996.
9. J.F. Allard and N. Atalla, Propagation of sound in Porous Media, Modelling Sound Absorbing Materials, Wiley, 2009.
10. B. Van den Nieuwenhof, G. Lielens, E. Rosseel, V. Dangla, C. Soize, M. Kassem and A. Mosson, "Optimal Design of the Acoustic Treatments Damping the Noise Radiated by a Turbofan Engine," in *American Institute of Aeronautics and Astronautics (AIAA)*, 2018.
11. L.N. Trefethen and D. Bau, Numerical Linear Algebra, Philadelphia: SIAM, 1997.

Robustness and Variability Prediction of Electric Machine Noise Using CAE

Mehdi Mehrgou(✉), Inigo Garcia de Madinabeitia Merino, Julian Pohn, Jorge Garmendia Gutierrez, Christoph Priestner, and Franz Zieher

AVL List GmbH, Hans-List-Platz 1, 8020 Graz, Austria
Mehdi.Mehrgou@avl.com

Abstract. Due to intensified development of electric vehicles, more and more focus is put on noise and vibration from the electric drive unit. One characteristic sound of electric vehicles is the tonal noise from the e-machine. This paper first shows how this tonal noise can be simulated and validated by measurement. It has been observed in vehicle measurements that identically designed electric axles can have a very high deviation in noise, so the question remains on how accurate the simulation results can predict this behavior. Using DoE methods and robust neural network (RNN) modelling, it is outlined how much manufacturing tolerances and material properties can affect the NVH behavior. This study concludes that even small deviations from the nominal geometries and material properties can have a significant effect on NVH. By applying the proposed method significant tolerances can be identified and revised in the early design stages of an electric motor.

Keywords: Electric motor · Automotive · NVH · Manufacturing tolerances · Material · DoE · Optimization · Forces · RNN · Robust neural networks

1 Introduction

In vehicles with conventional and hybrid powertrains, gear whine and electromagnetic noises are often not perceived because they are masked by the combustion noise. However, in battery electric vehicles (BEV) or plugin hybrids (PHEV) in electric-drive- modes, noises from gear whine, the e-motor and the inverter are easily audible because there is less background noise to mask these tonal noises.

The torque in electric motors is created by the combination of radial and tangential magneto-motive forces. These forces contain some higher order harmonics due to the characteristics of the e-motor, like the number of stator teeth, pole pairs or winding construction. These harmonics create losses, torque ripple and high frequency content of electromagnetic forces on the stator, which create vibration.

Regarding the electromagnetic forces, these are caused by reluctance changes along the magnetic circuit and are by definition normal to the surfaces. If the forces at the stator tooth-air gap interface are considered, the radial components are usually much bigger than the tangential forces. These forces are transferred to the housing

W. Siebenpfeiffer (Hrsg.): *Automotive Acoustics Conference 2019*, Proceedings, S. 200–210, 2020. https://doi.org/10.1007/978-3-658-27669-0_14

surface via the stator tooth, yoke, carrier and housing. However, both radial and tangential forces can contribute to NVH issues [1–4].

Beside the electromagnetic tonal noise, the gear and PWM tonal noise is also relevant in the application and discussed by the authors in publications [4, 6, 7]. Nevertheless, the focus of this paper is the electromagnetic noise and its deviation due to different manufacturing variations in geometry or material.

Any NVH issue has three aspects to be considered, which are the source, transfer path, and radiation surface. Numerous papers show the effect of eccentricity on the tonal noise of e-motors, but much less attention is paid to manufacturing tolerances of the laminate and its materials. Furthermore, the material properties of coils and glue materials are known to have a high effect on NVH and show high temperature-dependency as well [7]. Manufacturing tolerances will mainly affect the source while properties like fill factor or epoxy specifications will affect the stiffness and the transfer path of forces.

2 CAE Method for E-machine NVH

2.1 Electromagnetic Simulation

The first step towards NVH simulation is to create an electromagnetic FEM model (Fig. 1) of the electric machine(s) present in the electric drive unit (EDU). Analytical approaches are insufficient due to non-linearities of the system, such as electromagnetic saturation in the lamination or overmodulation in the inverter. Additionally, state-of-the-art control systems command the machine with different operation maps depending on temperature and non-linear dq-axis inductances. Fast and efficient electromagnetic FEM computing allows detailed characterization of electric machines accounting for these aspects.

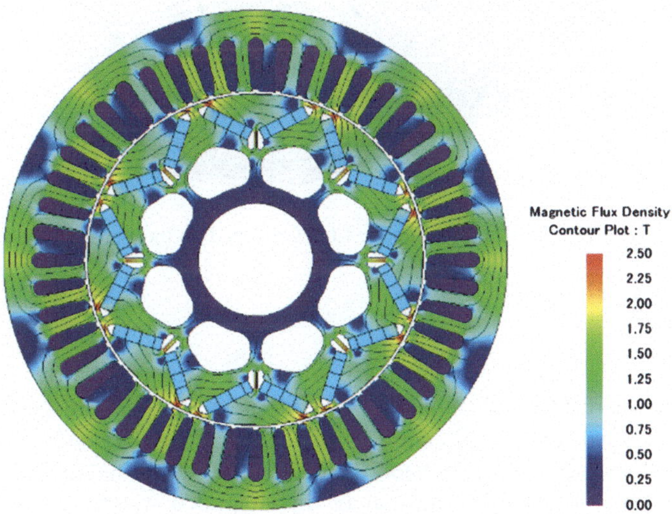

Fig. 1. Magnetic flux density plot from electromagnetic simulation of studied e-machine

These simulations do not only require reliable geometry and material data, but also the control inputs and inverter signals are needed. The application of conventional control algorithms like Maximum Torque Per Ampere (MTPA) and Field Weakening (FW) is common practice for permanent magnet e-machines which dominate the EDU applications. These control parameters also have some effect on the NVH behavior but are not discussed here.

Electromagnetic simulations are performed to obtain mainly torque, copper loss, iron loss, torque ripple and stator forces. More details about materials for magnetic simulation and boundary conditions and outputs like torque ripple or forces can be found in different publications [4].

2.2 NVH Simulation

Stator electromagnetic forces are mapped onto the NVH FEM model in frequency domain (Fig. 2). Material properties are orthotropic, with lamination and windings stiffer in the radial and axial directions, respectively. Strong variations of stiffness and damping coefficients due to temperature variations must be considered, as eigenfrequencies can shift by up to 30% and transfer functions might be modified, shown in [7].

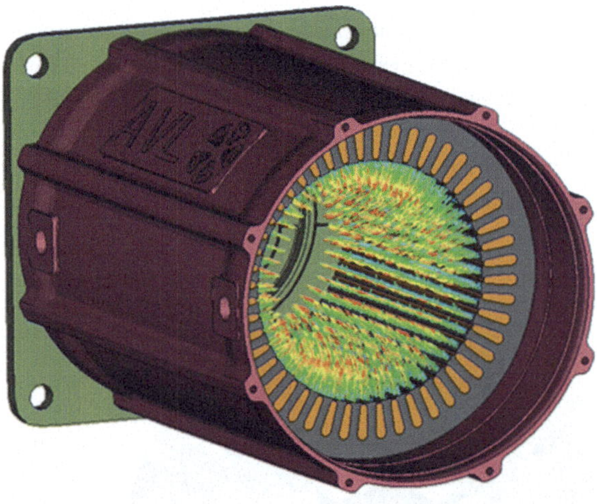

Fig. 2. Electromagnetic forces mapped on the stator

2.3 Validation

The authors have previously presented a simulation-measurement correlation of transfer functions, where material properties were derived by means of hammer tests in cold and warm conditions [7]. In addition, the effect of inverter harmonics on e-machine NVH issues has been correlated in [7, 8]. Once both material properties and inverter excitations are modelled accurately, the integrated system can be validated in operating condition. It is shown that the correlation accuracies are within 3 dB in the 48th order and around 1 dB in the frequency range of 1–6 kHz.

3 NVH Impact of Deviating Material Properties

3.1 Material Definition

In the NVH study of electric machines the correct characterization of the materials plays an important role. The main parameters are the Young's modulus (E), Poisson's ratio (υ) and the material density (ρ). However, when modelling the components, additional factors must be considered, such as the slot fill factor or the stator's and rotor's lamination factor. The first coefficient is the amount of copper filled in each slot, while the second one is related to the steel/epoxy ratio along the stack length. Paper [3] shows that materials like epoxy are viscoelastic, thus frequency-dependent and are characterized by high damping. It is relevant as well that the windings are properly modelled within the finite element analysis. Isotropic properties are no longer valid, so an anisotropic approach must be used. These additional factors make material modelling much more complex.

3.2 Identification of Material Properties

Since the e-machine assembly contains several materials and parts, the hardware hammer or tap tests are being used at different levels to identify the material properties using reverse engineering. This means that the material parameters are iterated in simulation to match the hammer test. The challenge becomes clear when hammer tests carried out on identically designed e-machine samples show different frequency responses, which indicates that different NVH behavior can be expected. Measurements can be done at different component levels: just the stator core, stator core with windings and at EDU level. The difference between two samples of identical designs from the same production line can be seen in Fig. 3.

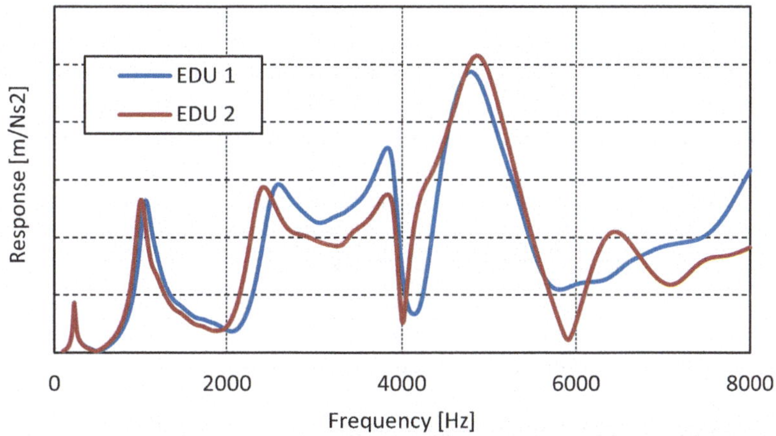

Fig. 3. Hammer test of two identical EDUs showing the difference in the response.

3.3 Structural Design of Experiments (DoE)

In the first DoE regarding the materials variation, the Young's Modulus of steel, copper and epoxy from both, windings and laminate, as well as the stack factor and the fill factor are taken as parameters. The ranges of these parameters in this case study were 10% for all Young's moduli, while for the stack and fill factor smaller deviations were considered. The range and values of the parameters should be selected based on the manufacturer specifications. The design of experiments consisted of 40 cases in which all parameters were varied within their mentioned upper and lower boundaries. As a result of the DoE study, the scatter band of the response due to the tolerances can be predicted (Fig. 4).

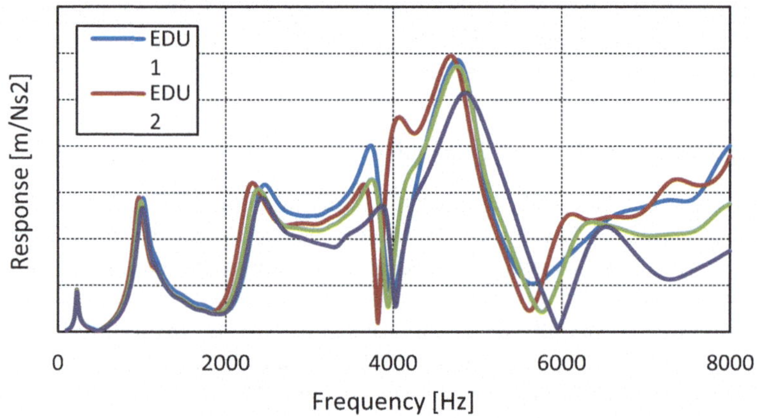

Fig. 4. Hammer test CAE simulation. Four sample cases from the DoE which show the scatter band of the response.

The DoE methodology enables a sensitivity analysis of all the material properties on the NVH response, which can help to identify the most significant parameters (stiffness, density, fill factor, etc.). This information is used to determine the most contributing parameter so that it can be specified as critical during the manufacturing process. In addition, the expected deviations caused by manufacturing can be covered with this method using CAE.

Using an AVL tool for postprocessing the DoE results, critical and trivial parameters can be easily identified. In the shown case study, most sensitive properties are the Young's modulus of the steel and the stack factor.

4 Sensitivity Analysis of Manufacturing Tolerances

4.1 Investigated Tolerances and Parameters

In the second DoE, the dimensional manufacturing tolerances due to stamping or laser-cutting are studied. An Internal Permanent Magnet (IPM) electric motor shown in Fig. 5 is chosen for demonstration. Tolerances are assigned to 15 geometrical parameters of the 2D topology. The main dimensions as well as the applied tolerances are given in Table 1. The tolerance range is based on the actual drawings which are strongly dependent on the manufacturing method. Therefore, a tolerance of ±0.1 mm is chosen as a baseline for these dimensions. Hereby, the rotor and stator outer and inner diameter which have a tolerance of ±0.2 mm are exceptions, as they represent the largest physical dimensions. In addition, an angular misalignment of ±1° of the V-shape permanent magnets is considered [5, 6].

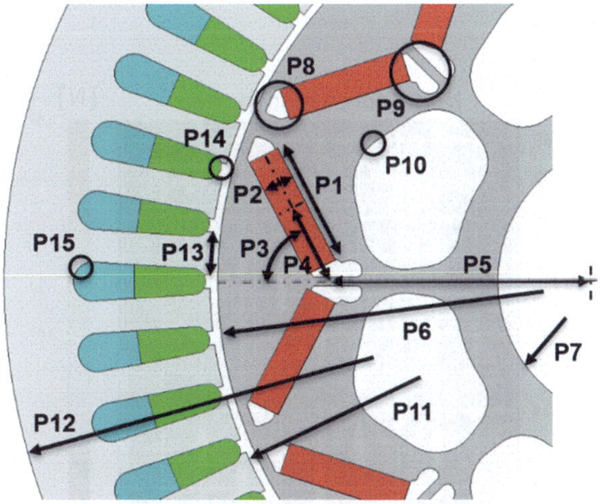

Fig. 5. Studied dimensions and tolerances of IPM topology

Table 1. Geometry parameters and applied tolerances (based on supplier data)

Number	Parameter	Tolerance	Unit
P1	Magnet width	±0.1	mm
P2	Magnet thickness	±0.1	mm
P3	Magnet angle	±1	deg
P4	Magnet position tangential	±0.1	mm
P5	Magnet position radial	±0.1	mm
P6	Rotor outer diameter	±0.2	mm
P7	Rotor inner diameter	±0.2	mm
P8	Tolerance outer flux barrier	±0.1	mm
P9	Tolerance inner flux barrier	±0.1	mm
P10	Tolerance cutout	±0.1	mm
P11	Stator inner diameter	±0.2	mm
P12	Stator outer diameter	±0.2	mm
P13	Stator tooth width	±0.1	mm
P14	Stator slot depth	±0.1	mm
P15	Coil stamping tolerance	±0.1	mm

4.2 Response Definition

The electromagnetic forces from the FEM calculation are used as response to see the effect on NVH. The characteristic force distribution in frequency as well as space domain of this e-machine which has 8 rotor poles and 48 stator slots is shown in Fig. 6. The 48th time order is the only time order to dynamically excite the 0th spatial order and is therefore the focus of the investigation. The NVH issue from the 48th time order is also measured in the testbed [7]. Furthermore, the effect of the studied tolerances on general performance characteristics like torque, efficiency and phase voltage is evaluated.

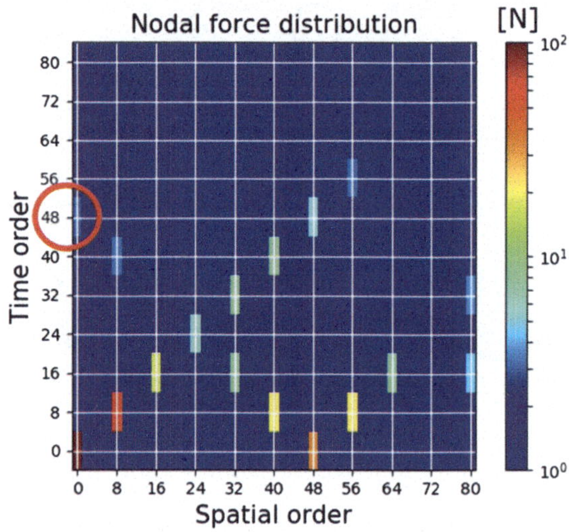

Fig. 6. Spatial order plot of the investigated IPM motor

4.3 DoE Method

To see the effect of the geometrical tolerances on the system responses, numerous simulations are needed. There are 15 geometrical variables and the impact of changes is considerably non-linear. Therefore, a proper DoE method is needed to reduce the number of necessary simulation cases to capture this behavior. The considered dimensions are split into 7 levels each, including the respective upper and lower tolerance limits. This resolution allows response detection of up to 6th polynomial order but excessively increases the amount of total possible tolerance combinations and hence simulation cases. To reduce the number of necessary simulation runs to an acceptable amount, the D-optimal approach is applied. A reduction to 1200 cases with a unique combination of tolerance values is achieved. The quality of the response surfaces is approved by prediction intervals (quality of simulated versus predicted responses).

4.4 Sensitivity of Parameters and Effect on NVH

Since the electromagnetic forces vary across the speed range, multiple operating points have to be simulated. The parameters and responses of all cases are used to model the system behavior with the aid of robust neural network (RNN) methods. Figure 7 shows the sensitivity of torque and the radial stator forces of 48th time order in three different operating points spread across the speed range. The torque is mainly sensitive to the tolerances of magnet angle, air gap and the stator slot stamping tolerance. The torque sensitivity concerning the other parameters is small and does not change in different operating points. As can be seen in Fig. 7, this behavior is very much different for the radial stator forces of 48th time order relevant for NVH. The sensitivity of this force components changes strongly between operating points and even inverse characteristics for some tolerances in different operating points can be seen. Figure 7 shows the system sensitivities for the nominal parameter values. However, the different parameters affect each other. If one is changed, the characteristics of the others are likely to change as well.

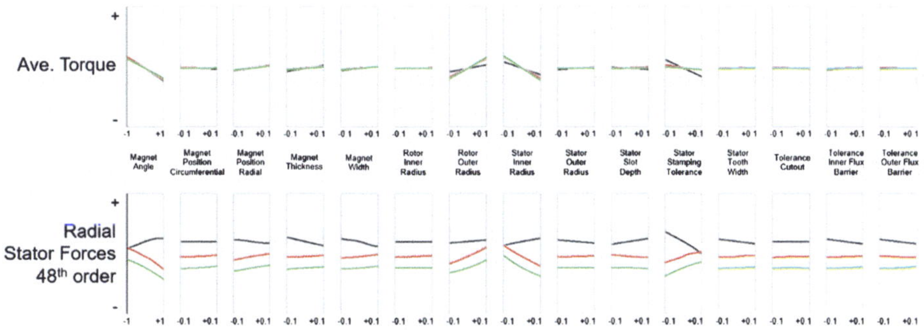

Fig. 7. Sensitivities of torque and 48th order stator forces in three operating points: 2500 rpm (black), 5000 rpm (red), 7500 rpm (green)

Due to the high number of variables, their non-linear behavior and mutual influence, only DoE can provide an efficient way of studying these tolerances. The blue dots in Fig. 8 show the radial and tangential 48th order stator forces for all the cases defined by DoE in one operating point. Although just these 1200 cases have been simulated in FEM tool, modelling the system behavior with the aid of neural networks can predict the outcome for hundreds of thousands or even millions of other tolerance combinations (illustrated as orange cloud in Fig. 8). By applying numerical optimization, the mathematical system model is used to find the best and worst cases concerning NVH (highest and lowest forces) which are shown by the green pareto fronts in Fig. 8. The minimum and maximum cases for the radial stator forces of 48th time order differ by 13 dB. These two variants are selected for the NVH simulation.

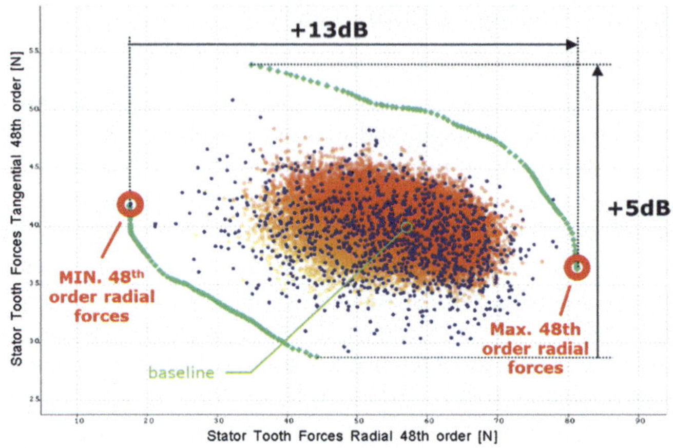

Fig. 8. DoE FEM results (blue dots) and mathematical system model (orange cloud) of the 48th time order forces including the pareto fronts (green lines)

For NVH evaluation, the normal surface velocity levels are evaluated. To illustrate the phenomenon across the whole operating range of the e-motor the structural FE results are interpolated between multiple operating points. The surface integral of the velocities across the speed range is shown in Fig. 9.

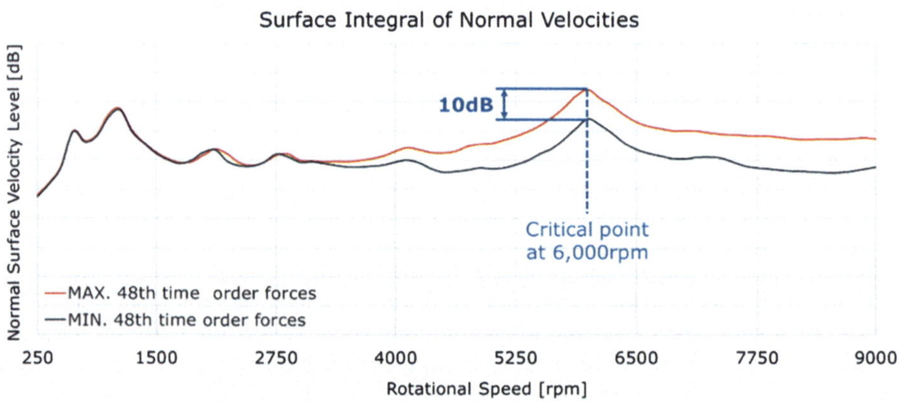

Fig. 9. Surface velocity integral across operating range

A global maximum and therefore critical operating point for NVH can be seen at 6000 rpm. In addition, a difference in surface velocity levels of up to 10 dB is highlighted at this point – between best and worst tolerance combination. It can be noticed that a difference in NVH behavior starts to show at around 3000 rpm. At this speed the field-weakening area of the e-machine starts. The detailed illustration of the surface velocities at 6000 rpm is shown in Fig. 10. Around this frequency the stator breathing eigenmode (0th spatial order) is excited by the 48th time order of the stator forces. Airborne noise simulations prove the 10 dB difference in sound pressure levels as well [6].

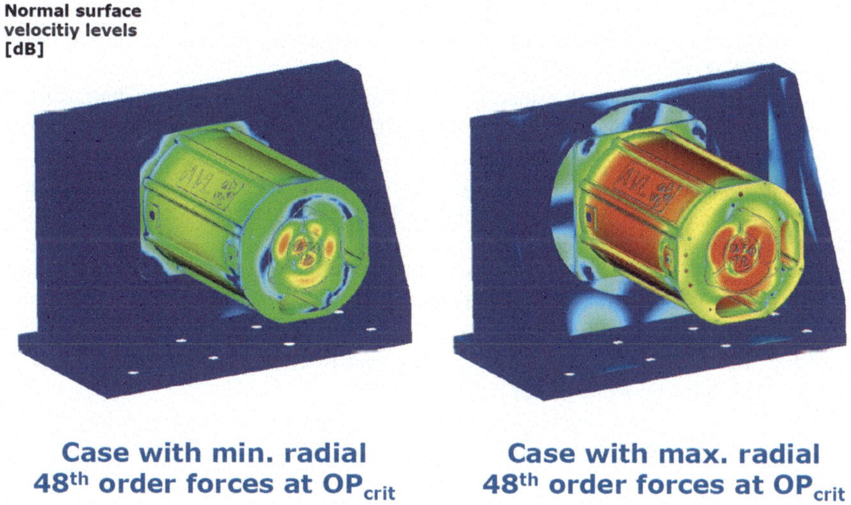

Fig. 10. Surface velocity levels at 6000 rpm and maximum torque

5 Conclusion

Within a several EDUs with identical design, significant differences in NVH behavior have been observed, which raises the question of the root cause for these deviations. In this paper, both electromagnetic and structural FEM simulation models and DoE are used to study the effect of material and manufacturing tolerances on NVH.

It is shown that different material properties can lead to considerable deviations in NVH responses. Obtaining the proper material data for the simulation is a complex process that requires hammer testing of hardware.

On the other hand, geometrical tolerances have a very high impact on the NVH responses of electric machines. It is highlighted that only DoE and optimization methods provide an efficient way to detect best and worst-case tolerance combinations. The resulting NVH behavior is assessed by comparing the surface velocity levels which show a deviation of up to 10 dB due to the dimensional tolerances.

The findings of this investigation can support the development of e-machine designs which are more robust to tolerances and can guide the manufacturing of EDUs by identifying critical parameters needed to be controlled during production and explain the deviation of simulation to measurement deviations.

References

1. Pyrhonen J. et al.: Design of Rotating Electrical Machines. 2nd ed. John Wiley & Sons, 2014. ISBN: 978-1-118-58157-5.
2. Dupont J., Aydoun R. and Bouvet P.: Multiphysics Modelling to Simulate the Noise of an Automotive Electric Motor. In: SAE Technical Paper 2012-01-1520 (2012).
3. Pubadubsin et al.: Mechanical Material Properties for Structural Simulation Model of Switched Reluctance Machines. In: 2016 XXII International Conference on Electrical Machines (ICEM) (2016), pp. 2293–2299.
4. Mehrgou, M., Zieher, F., and Priestner, C., "NVH and Acoustics Analysis Solutions for Electric Drives," SAE Technical Paper 2016-01-1802, 2016, https://doi.org/10.4271/2016-01-1802.
5. N. Taran et al.: "On the effect of design tolerances on the performance of synchronous PM machines evaluated according to the IEEE Std 1812". In: 2017 IEEE International Electric Machines and Drives Conference, IEMDC 2017 (2017), pp. 1–7.
6. Pohn, J.: FEM Based Analysis of Manufacturing Tolerances and Eccentricity in Electric Motor, Master's thesis, Graz University of Technology, 2019.
7. Mehrgou, M., Garcia de Madinabeitia, I., Graf, B., Zieher, F. et al., "NVH Aspects of Electric Drives-Integration of Electric Machine, Gearbox and Inverter," SAE Technical Paper 2018-01-1556, 2018, https://doi.org/10.4271/2018-01-1556.
8. Mehrgou, M., Garcia de Madinabeitia, et al, "Simulation and Validation of Inverter-related NVH Issues in Electric Vehicle Drives," ATZ Extra, 01.10.2018, Automotive Acoustics, Sonderheft 2/2018.

Development of a Methodology for Acoustic Analysis and Optimization of Mechatronic Fuel Injection Systems

Christian Mühlbauer[1(✉)], Hans-Peter Rabl[1], and Herrmann Rottengruber[2]

[1] CHP Messtechnik GmbH, Franz-Mayer-Straße 1, 93053 Regensburg, Germany
christian.muehlbauer@chp-messtechnik.de

[2] OvGU Magdeburg, Lehrstuhl Energiewandlungssysteme für mobile Anwendungen, Universitätsplatz 2, 39106 Magdeburg, Germany

Abstract. To identify the sound quality of fuel injection systems as a front-loading measure, a system test bench has been developed that simulates the structure and airborne noise emission. The setup of the system test bench includes an assembly of the acoustical relevant engine parts and an encapsulated drive for the high-pressure pump. A specific control unit enables a high variability for the operation strategy of the injection system.

The focus of this study is on the development of a methodology to analyze and optimize the sound quality of gasoline direct injectors in critical operation points by suitable active means. For this, the injection parameters in engine idle have been identified and subsequently transferred to the system test bench. Additionally, the injection parameters have been optimized acoustically and the impacts on the key functions regarding injection rating and shot-to-shot scattering of the injected fuel mass are discussed.

Keywords: Injection system · Ticking noise · System test bench

1 Introduction

The tightening of the emission legislation as well as the political and social demand on sustainable mobility forces the automotive industry to develop complex high precision mechatronic powertrain systems. The increasing precision of the mechatronic systems usually leads to an increasing excitation of the structure and therefore to an increase in noise emission. One of these mechatronic powertrain systems is the gasoline direct injection (GDI), depicted in Fig. 1. The GDI is increasingly replacing the multi-point injection (MPI) also in the volume market [1].

By means of the GDI in combination with exhaust turbocharging and variable camshaft adjustment it is possible to reduce the fuel consumption while simultaneously increasing engine torque and power. Several degrees of freedom regarding injection strategy, as for example fuel pressure, quantity and crankshaft position of the

W. Siebenpfeiffer (Hrsg.): *Automotive Acoustics Conference 2019*, Proceedings, S. 211–225, 2020. https://doi.org/10.1007/978-3-658-27669-0_15

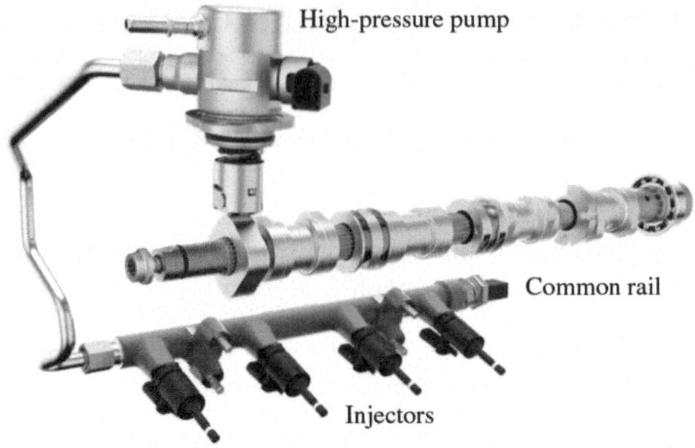

Fig. 1. Main system components of a gasoline direct injection [3].

injections, enable a reduction of the raw emissions. Furthermore, it allows to shorten the emission-critical engine warm up phase [2].

1.1 Approach and Content of the Paper

NVH-Optimization of GDI systems by active means could be a suitable measure to reduce the emitted sound pressure and to improve the perceived sound quality. The test benches and equipment needed for the tests are described in Sect. 2. In order to be able to develop suitable active measures, Sect. 3 deals with the basic noise generation mechanisms of GDI injectors. In Sect. 4 two NVH measures are discussed on their acoustical and functional effects. Section 5 focuses on the improvements of the perceived sound quality by analyzing various psychoacoustic parameters.

2 System Test Bench Configuration

The sound quality of a vehicle is primarily influenced by the perceived noise of the powertrain. Particular noise sources that impair the sound quality need to be detected and eliminated in the development process. One main problem in the NVH development is the noise source identification and characterization. Besides the classic NVH analysis methods, such as bandpass-filtering and temporal windowing, there is a potential approach that is based on the blind source separation [4–7].

To ensure a reliable separation of the noise emitting components, the classification algorithm needs a complex preconditioning. If there are major changes in function and structure of an engine component during the development process, certain classification criteria, such as sound pressure level (SPL) and dominating frequency range, could change as well. For such applications a system test bench has been developed that enables a direct and isolated analysis of GDI components.

2.1 System Test Bench Setup

The setup of the NVH system test bench basically comprises a dragged and unfired partial engine (see Fig. 2). As shown in Fig. 1, the single-piston high-pressure pump is driven by a multiple lobe that is integrated on the camshaft. To realize the circulation of the injected fuel, the crankcase is replaced by a crankcase dummy without crank drive. Therefore, the camshaft is directly driven by an asynchronous motor. The emitted noise of the asynchronous motor and its frequency converter is insulated by a three-layered encapsulation.

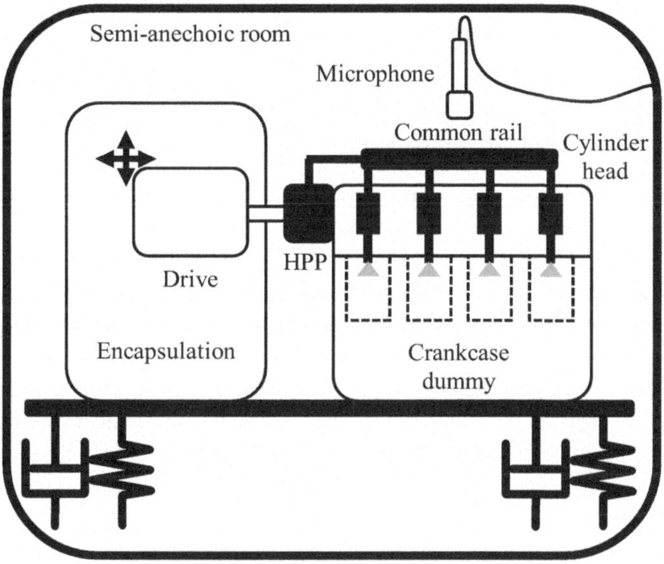

Fig. 2. Setup of the NVH system test bench.

In order to test the NVH measures with regard to the key functions of the injection system, such as injection rate, injected fuel mass and shot-to-shot scattering, the tests are also carried out on an injection rate test bench.

2.2 System Test Bench Control and Measurement

The injection system is controlled by a development ECU with several degrees of freedom regarding the injection strategy. This enables the flashing of different current profiles of the injector and high-pressure pump control in a time- and cost-efficient way. The measurements are performed in a semi-anechoic room with a lower cut-off frequency of 150 Hz. The desired frequency range for the acoustic analyses is between 500 Hz and 20 kHz. The required signals are recorded with a multi-channel data acquisition system. Four ½″ free field condenser microphones in 1 m distance to the noise source are used to measure the airborne sound. The position of the microphones has been varied to figure out the critical direction of the noise radiation. Furthermore, the control current of the injectors and high-pressure pump, the high and low resolution rotational speed of the camshaft as well as various measuring points for temperature and pressure are acquired and recorded.

The emitted noise of the GDI is mostly perceived as annoying in engine idle and in the lower speed and load range [1]. In that operation points the masking of other noise sources, such as combustion, wind and tyres, is low. Therefore, the electrical control of the GDI was measured on an engine test bench and transferred to the system test bench.

2.3 Preprocessing of Measurement Data

The recorded measurement data with a duration of 20 s is further processed in MATLAB. The microphone measurements are bandpass-filtered between 250 Hz and 50 kHz. Since only the injector noise is examined in this study, the emitted noise of the high-pressure pump is removed via a temporal windowing in the time signal. For this purpose, the injectors and the high-pressure pump are controlled in phase opposition. The correct temporal windowing is realized by a peak analysis in the control signals of the injectors and high-pressure pump. In addition, the time lag due to the opening delay as well as the structure-borne and airborne sound propagation time is taken into account. The bandpass-filtered and separated microphone signal can then be used for further analyses, such as the calculation of third-octave bands and psychoacoustic parameters.

For comparison purposes the wavelet analysis uses single representative control cycles. The time signal is first sorted cyclically and injector-specific. After calculating the injector-specific cyclic mean of the microphone signal, the control cycle which has the smallest deviation from the mean is used as a representative control cycle. Subsequently, the injector which has the highest RMS value and therefore the highest noise emission is used for the wavelet analysis.

2.4 Validation of the System Test Bench Setup

In order to validate the experimental setup of the system test bench, reference measurements on an identical full engine were performed in the semi-anechoic room. This was in particular important for the injectors because of their lateral position. Their noise radiation is primarily affected by reflection and structure-borne noise transmission from surrounding components.

Therefore, the injectors were controlled with the in Sect. 2.2 mentioned development ECU with internal signal generation at engine standstill and without fuel injection. That approach corresponds to an applied structure excitation. Thereupon, the tests were identically repeated on the partial engine of the system test bench. At first, the partial engine of the system test bench was only equipped with the injection system and the cylinder head. Subsequent, further engine parts, such as the intake manifold and the crankcase ventilation, were attached successively until the radiated noise spectrum matched with the reference within a desired tolerance of 5 dB per third-octave band.

Figure 3 shows a comparison of the third-octave spectrum between the full engine and the system test bench with the attached acoustical relevant engine parts. Apart from the third-octave bands with center frequencies of 0.8 kHz and 1 kHz, the spectral distribution shows a similar trend and is within the desired tolerance of 5 dB per third-octave band.

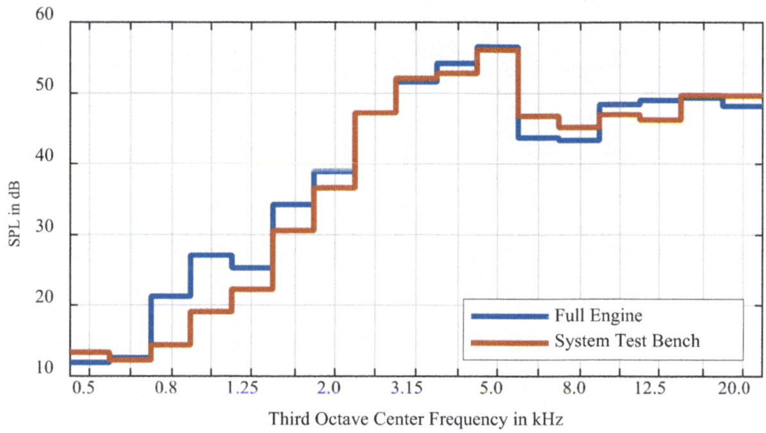

Fig. 3. Comparison of the noise spectrum between full engine and system test bench.

3 Mechatronic Noise Sources of GDI Injectors

Especially the ever increasing injection pressures and the faster actuation times lead to a greater energy input and change the NVH of the GDI system. Together with the stricter noise emission limits, modern GDI systems are also developed with consideration of the acoustic performance.

In the further course, only the acoustics of solenoid injectors will be discussed. The perceived injector noise is elicited by its opening and closing mechanism. Provoked by the fast actuation times, the valve needle and armature is exposed to large accelerations. The braking of the valve needle and armature when the stop is reached results in a structural excitation. The high density of energy, both locally and temporally, leads to an impulsive and high frequency noise. Figure 4 depicts the fundamental noise sources of a GDI injector. Due to various technological measures taken by the manufacturers of injection systems, such as the decoupling of armature and needle or so-called idle stroke concepts, individual noise generation mechanisms may have to be reconsidered under certain circumstances [8].

Ticking is particularly critical for gasoline engines, since in contrast to diesel engines the combustion noise has a relatively low sound pressure level [5]. Another problem is the decrease of the average engine noise with increasing frequency. This means that the high frequency part of the ticking is perceived more clearly, even if there is a lower sound pressure level compared to the rest of the engine noise [9].

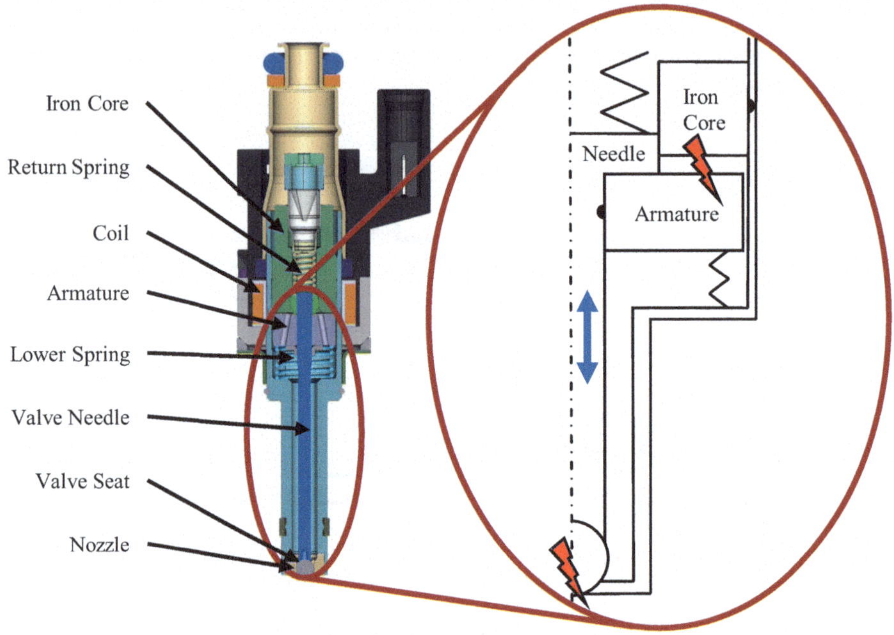

Fig. 4. Setup and fundamental noise sources of a GDI solenoid injector [8].

Figure 5 depicts the control current of an injector (bottom) and a wavelet analysis of the resulting microphone signal (top) with a fuel pressure of 100 bar. The wavelet analysis shows a high-frequency and impulsive noise emission with a dominating frequency range of about 15 kHz with a time delay of about 4 ms to the electrical control start.

In addition to the airborne sound propagation time, the time delay is due to the hydraulic opening delay and the structure-borne sound propagation time. Analogous to the electrical control time of 3 ms, the wavelet analysis shows a further pronounced impulsive and high frequency sound emission after 3 ms. These two local sound pressure maxima are caused on the one hand by the opening bounce (armature – iron core) and on the other hand by the closing bounce (valve needle – sealing seat) of the injector. On the in this study investigated injector the closing bounce has a slightly greater acoustic influence than the opening bounce.

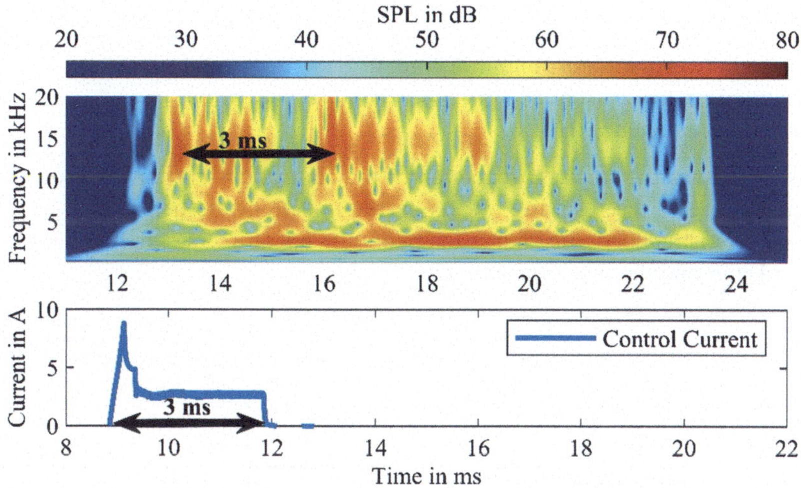

Fig. 5. Wavelet analysis of an injection process with 100 bar fuel pressure.

4 Approaches to Noise Reduction

4.1 Reduction of Control Current

To optimize the acoustic performance of GDI injectors, the opening bounce can be reduced. The resulting force F_{res} with which the valve needle and armature are pressed against the iron core is calculated according to Eq. (1):

$$F_{res} = F_{mag} - F_{spring} - F_{fluid} \quad (1)$$

where

F_{res} = Resulting opening force, N
F_{mag} = Magnetic force, N
F_{spring} = Return spring force, N
F_{fluid} = Fluid force, N

Since the force of the return spring cannot be influenced and the fluid force depends on the desired injection pressure, the magnetic force is lowered to reduce the resulting force. Equation (2) shows the calculation of the magnetic force as a function of the gap width between armature and iron core and the control current with the simplified assumption of stationary conditions [10]:

$$F_{mag}\left(x_{gap}, I\right) = \frac{1}{2} I^2 \frac{dL}{dx_{gap}} \quad (2)$$

where

x_{gap} = Gap between armature and iron core, m
I = Electrical control current, A
L = Electrical inductance, H

Figure 6 shows the stepwise reduction of the control current for the acoustic optimization above and the resulting effects on the injection rate below. The further the control current is reduced, the greater the hydraulic opening delay between the electrical control start and the injection start gets.

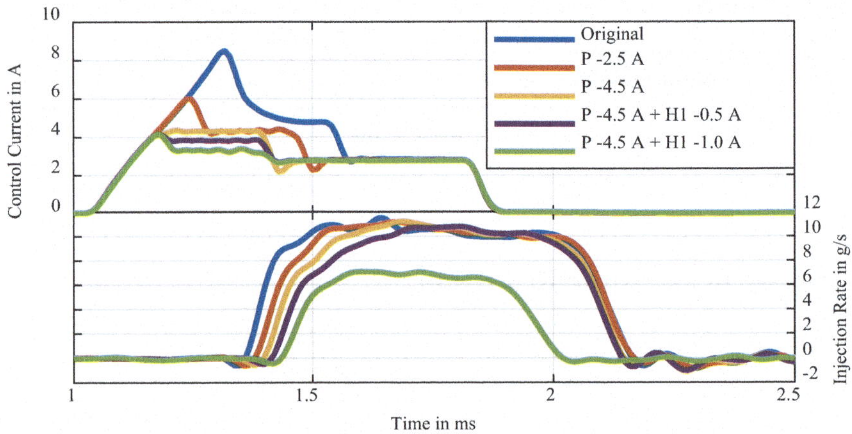

Fig. 6. Stepwise reduction of the control current (above) and effects on the injection rate (below).

In addition, as the control current is reduced, the gradient of the injection rate becomes flatter, indicating a reduction in the opening speed of the valve needle and armature. In the green curve with a peak reduction of 4.5 A and a Hold1 phase reduction of 1.0 A, the control current has been reduced to an extent at which the injector no longer opens completely.

This so-called ballistic partial stroke operation has a particularly negative effect on the shot-to-shot scattering between the individual injections, shown as standard deviation in Table 1.

Although the mean value of the injected fuel mass is also reduced, this effect can be corrected by extending the electrical control time. Since the control profile with a reduction of the peak by 4.5 A and the Hold1 phase by 0.5 A still shows satisfactory results regarding opening dynamics and shot-to-shot scattering, this control profile was selected for the acoustic optimization.

Table 1. Mean and standard deviation of the injected fuel mass over 100 injections.

Control current profile	Mean (mg)	Standard deviation (%)
Original	7.188	Basis
P −2.5 A	6.950	+1.493
P −4.5 A	6.502	+11.940
P −4.5 A + H1 −0.5 A	6.067	+4.478
P −4.5 A + H1 −1.0 A	3.179	+271.642

Figure 7 depicts a comparison of the third-octave spectra of the original and current-reduced control profile. The third-octave SPL can be reduced by up to 7 dB starting at a third-octave center frequency of ≥1.6 kHz. In contrast, the two third-octave levels with the third-octave center frequencies 2.5 kHz and 3.15 kHz appear conspicuous. They show a significant increase in SPL compared to the original current profile.

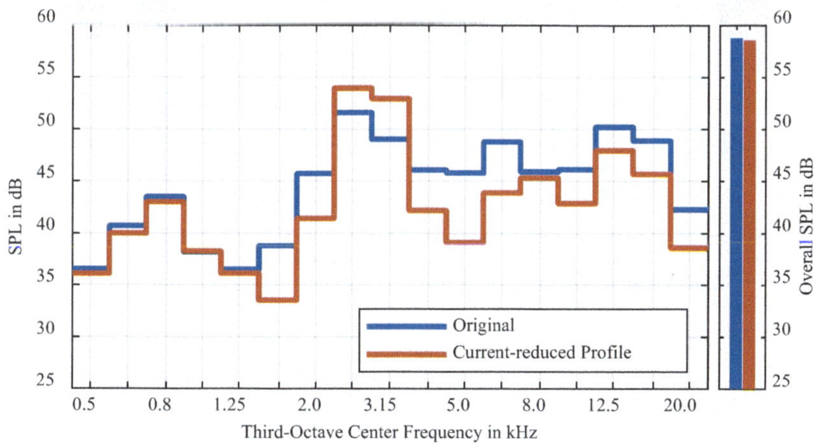

Fig. 7. Comparison of the third-octave spectrum of original and current-reduced control profile.

In order to comprehend the reason of this increase in SPL, a wavelet analysis was performed (see Fig. 8). It can be seen that the high-frequency closing bounce creates another distinct noise phenomenon for the above-mentioned critical third-octave center frequencies 2.5 kHz and 3.15 kHz.

Considering the injection rate of both control profiles in Fig. 9, an earlier end of injection can be seen with the current-reduced control profile, even though the control duration is the same. Furthermore, there is a greater fluctuation amplitude after the injection end which indicates an increased closing bounce of the valve needle against its sealing seat. This effect is probably due to the reduced residual magnetization during the closing process caused by the lower control current. According to Eq. (1), the resulting return force increases.

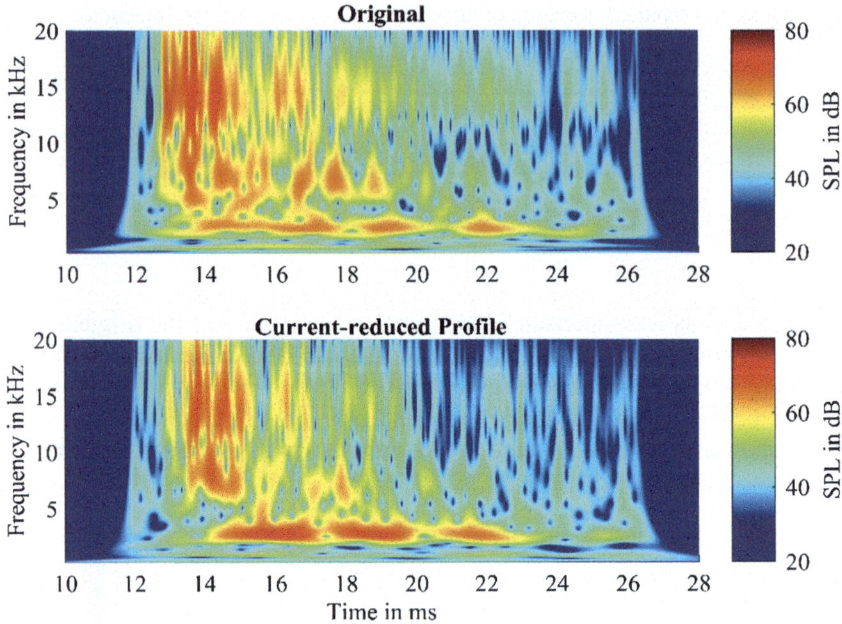

Fig. 8. Comparison of the wavelet analysis of original and current-reduced control profile.

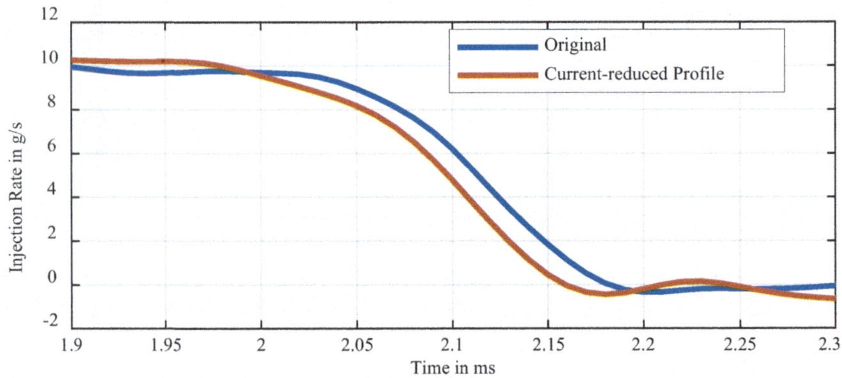

Fig. 9. Comparison of the injection rate of original and current-reduced control profile.

4.2 Softlanding

In order to compensate the closing bounce, an additional braking current is applied at the end of the electrical control (see Fig. 10 above). This increases the magnetic force during the closing process and thus reduces the resulting force which pushes the valve needle back into its sealing seat.

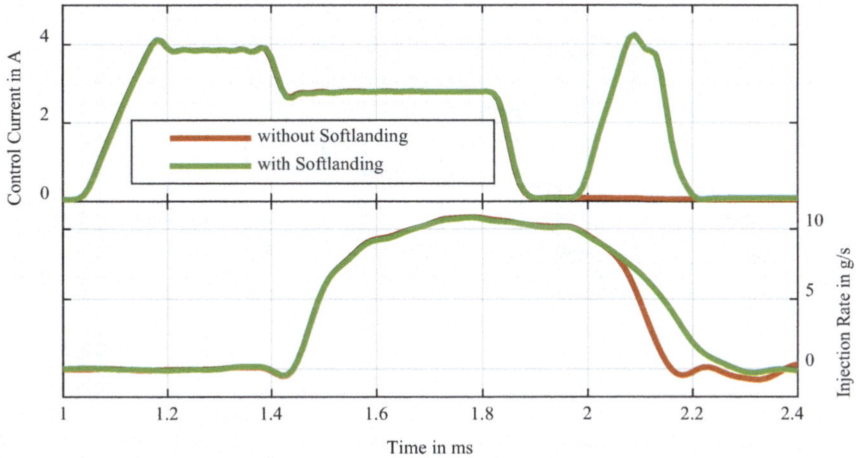

Fig. 10. Comparison of control current (above) and injection rate (below) of current-reduced control profile with and without softlanding

The selection of the correct separation and control time of the brake pulse was determined experimentally on the NVH and injection rate test bench. If the mentioned parameters are set incorrectly, the injection can be mistuned significantly on the one hand. On the other hand, no NVH benefits can be achieved.

The lower diagram in Fig. 10 indicates the reduced closing dynamics of the valve needle due to the flatter injection rate curve during the closing process by means of the brake current. In addition, the fluctuation amplitude of the injection rate at the end of the injection process could be reduced.

However, the measure indicates the disadvantage of an increasing shot-to-shot scattering of the injected fuel mass. Thus, the increase of the standard deviation of +220.896% with softlanding is considerably higher than the increase of +4.478% without softlanding. Further investigations revealed that a fluctuation in the separation time is responsible for the high shot-to-shot scattering. This is caused by the development ECU and can possibly be reduced if the braking current is not applied as a second short injection pulse but integrated into the current profile itself.

In Fig. 11, the measure is considered acoustically. The third-octave SPLs can be reduced by up to 9 dB compared to the original control profile at third-octave center frequencies $\geq$1.6 kHz. The overall SPL was reduced by approx. 4.5 dB.

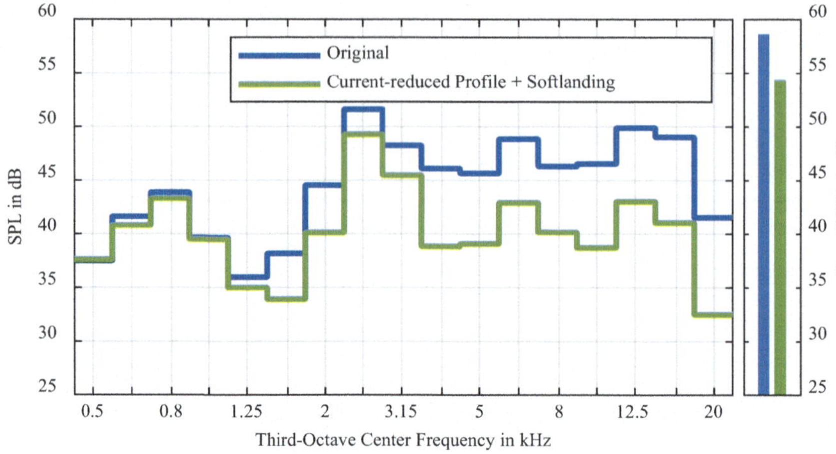

Fig. 11. Comparison of the third-octave spectrum of original and current-reduced control profile with softlanding.

5 Psychoacoustic Analysis

The auditory perception is one of the sensations that is used to evaluate the quality of a product. By analyzing the SPL of a particular sound, no statement about the actual hearing sensations can be made. To go beyond the physical consideration of sounds, a relationship between stimuli and its auditory perception has to be made. Psychoacoustics deals with this relationship. For this purpose, psychoacoustic models have been set up and evaluated with hearing tests.

The auditory perception of sounds can be split into several hearing sensations. Unfortunately, only two of the hearing sensations are standardized so far. On the one hand the psychoacoustic metric loudness, which accounts for the subjectively perceived volume of a sound [11–13]. One the other hand sharpness, which can be described as the ratio of specific loudnesses of higher frequency content to total loudness [14].

Figure 12 depicts a comparison of the loudness (left) and sharpness (right) of the original, the current-reduced and the current-reduced profile with softlanding (SL). The loudness is, as usual, represented as 5% percentiles, i.e. 5% of all calculated values are above this value. The sharpness is displayed as the mean of all calculated values. As already expected from the analysis of the SPLs and third-octave spectra, the loudness shows a significant reduction with the application of softlanding. The sharpness, however, shows only a small linear decrease with increasing optimization effort.

The perceived noise of an injector is usually called ticking and is characterized by a high impulsiveness. The question arises whether the two standardized psychoacoustic parameters are sufficient to adequately describe the sound quality of an injector. Noise impulses show short-term changes in time in the sound pressure signal. For the analysis of such signals, various methods for the determination of impulsiveness have been established.

A simple approach to impulsiveness is the kurtosis. It results from the temporal change of the signal amplitude over a given integration time. SOTTEK describes an

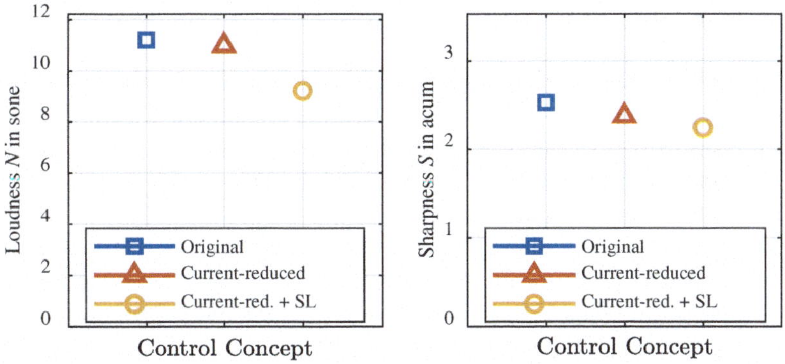

Fig. 12. Comparison of the loudness (left) and sharpness (right) of the original, the current-reduced and the current-reduced profile with softlanding (SL).

alternative method for calculating the impulsiveness. This is based on a model which combines the physiological properties of the auditory system. The perceived impulsiveness depends on the pulse frequency and increases up to frequencies of 10 Hz. Above this frequency, the sound impression changes into a roughness sensation. The impulsiveness also increases with increasing signal level and with increasing edge steepness of the impulses [15].

Since the variation of the control profiles was always performed at the same idle speed and single injection, the pulse frequency remained constant. Thus, changes in the calculated impulsiveness were caused by varying signal levels and edge steepnesses of the ticking noises. Figure 13 shows a comparison of the kurtosis (left) and impulsiveness according to SOTTEK (right) of the original, the current-reduced and the current-reduced profile with softlanding (SL). In contrast to the impulsiveness according to SOTTEK, the kurtosis calculates a slightly higher value for the current-reduced profile than for the original current profile. The current-reduced profile with softlanding shows the lowest value in both calculation approaches.

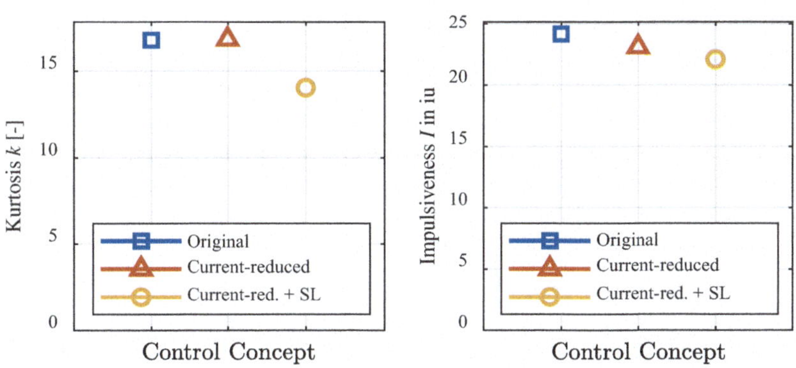

Fig. 13. Comparison of the kurtosis (left) and impulsiveness according to SOTTEK (right) of the original, the current-reduced and the current-reduced profile with softlanding (SL).

6 Summary and Outlook

For the acoustic analysis and optimization of GDI systems, a system test bench has been developed which allows a direct and isolated sound-based testing. The system test bench essentially consists of a dragged and unfired partial engine. The injection system is controlled by a development ECU which allows many degrees of freedom with regard to the injection strategy.

By analyzing the noise sources of a GDI injector, it turned out that the greatest excitation is caused by its opening and closing bounce. In the further course of the study, an approach was made to reduce these excitation sources by means of control engineering measures. The opening bounce could be reduced by a current-reduced control of the injector with satisfactory shot-to-shot scattering. Due to the lower residual magnetization during the closing process the perceivable closing bounce increased, what led to an increase in SPL in individual third-octave bands.

In order to take this effect into account, a second pulse was applied as braking current at the end of the electrical control. If the correct control parameters are selected, this softlanding function shows considerable improvements in SPL, both spectrally and overall. However, a slight cyclical fluctuation of the separation time between the two control pulses leads to a significant increase in the shot-to-shot scattering.

In order to reduce the increased shot-to-shot scattering in the softlanding function, further measures regarding the electrical control have to be developed. The fluctuation of the separation time can possibly be reduced if the braking current is already integrated into the actual control profile and not implemented as a second short control pulse as it was previously the case.

In the psychoacoustic analyses, the current-reduced profile with softlanding shows the lowest values for the standardized metrics loudness and sharpness as well as for the two described calculation methods for impulsiveness. However, a direct statement regarding the customer-relevant noise quality cannot be made. This requires a validated model of the perceived annoyance, taking impulsiveness into account.

Since modern high-pressure pumps are also a perceptible noise source due to digital control, possibilities for active noise reductions for this mechatronic GDI component have to be developed as well. Therefore, the test software of the development ECU has already been extended in order to investigate various control strategies with regard to acoustic benefits.

Furthermore, the effects of the NVH measures in engine operation must be considered. On the one hand, how the measures affect acoustically and whether they may become ineffective due to possible masking effects of other noise sources. On the other hand, the influence on key functions such as smooth running, exhaust emissions and fuel consumption needs to be examined.

References

1. Braun, C. W. et al.: Acoustic Noise Assessment of Gasoline Direct Injection Fuel Injectors Using Taguchi Methods, SAE Technical Paper 2011-01-1216, 2011.

2. Merker, G. P.; Teichmann, R.: Grundlagen Verbrennungsmotoren, 8. Edition, Wiesbaden: Springer Vieweg, 2018.
3. Eichler, F. et al.: Der neue EA211 TSI®evo von Volkswagen, 37. Internationales Wiener Motorensymposium, 2016.
4. Lee, D.D.; Seung, H.S.: Algorithms for Non-Negative Matrix Factorization, Neural Information Processing Systems, 2000.
5. Atzler, M.: Bewertungsmodelle für impulsartige Störgeräusche von Verbrennungsmotoren, Dissertation, RWTH Aachen, 2011.
6. Selle, A.: Motorgeräuschkomponenten, FVV Final Report 1092, 2014.
7. Schumann, C.: Motorgeräuschkomponenten II, FVV Final Report 1207, 2018.
8. Frenzel, H. et al.: Direkteinspritzsysteme für Ottomotoren zur Erfüllung der Real Driving Emission, 10. Tagung Diesel- und Benzindirekteinspritzung, 2016.
9. Zeng, P.; Solferino, V.; Stickler, M.: Direct Injection Design Principles for Noise Vibration Harshness, SAE Technical Paper 2017-01-1052, 2017.
10. AVL BOOST Hydsim Userguide, 2013.
11. Genuit, K.: Sound-Engineering im Automobilbereich. Berlin, Heidelberg: Springer, 2010.
12. DIN 45631: Berechnung des Lautstärkepegels und der Lautheit aus dem Geräuschspektrum – Verfahren nach E. Zwicker, Berlin: Beuth Verlag, 1991.
13. DIN 45631/A1: Berechnung des Lautstärkepegels und der Lautheit aus dem Geräuschspektrum – Verfahren nach E. Zwicker – Änderung 1: Berechnung der Lautheit zeitvarianter Geräusche, Berlin: Beuth Verlag, 2010.
14. DIN 45692: Messtechnische Simulation der Hörempfindung Schärfe, Berlin: Beuth Verlag, 2009.
15. Sottek, R.; Vranken, P.; Busch, G.: Ein Modell zur Berechnung der Impulshaltigkeit, DAGA, 1995.

Advanced NVH Measurement Data Processing of Electric Motors under Electromagnetic Excitations

K. Degrendele(✉), E. Devillers, and J. Le Besnerais

EOMYS ENGINEERING, 121 rue de Chanzy – Hellemmes, CS 10128, 59030 Lille Cedex, France
contact@eomys.com

Abstract. This paper presents several Noise, Vibration, Harshness (NVH) post-processing techniques used on a dynamic acquisition system to troubleshoot acoustic noise and vibration issues due to electromagnetic forces in e-powertrains. Standard NVH post processing techniques based on Order Tracking (OT), Operation Deflection Shapes (ODS) and Experimental Modal Analysis (EMA) are presented and as well as their application to discriminate the frequencies of magnetic forces responsible for NVH issues. However, they do not filter the different wave-numbers of vibration waves. Some special NVH post processing techniques are introduced to go deeper in the analysis and quickly capture electromagnetically-excited NVH root cause. They are based either on "spatiogram" or on flux sensor waveform post processing. "Spatiogram" relies on the Discrete Fournier Transformation of accelerations measured on the machine. It aims at introducing a space information besides the standard time and frequency information. It can be seen as an extension of the spectrograms done at variable speed as it efficiently discriminates wavenumbers of magnetic forces responsible for NVH issues. Experimental measurements run on a Permanent Magnet Synchronous Motor (PMSM) are interpreted and confirmed by simulation results. Similarly flux sensor waveforms can be post-processed to quantify geometrical and magnetic asymmetries, to estimate radial magnetic forces per tooth and derive Frequency Response Functions (FRF) of stator teeth under radial electromagnetic excitations.

Keywords: Electric powertrain · NVH · Testing

1 Introduction

Electromagnetic forces in e-powertrains are a source of so-called electromagnetically-excited acoustic noise, or electromagnetic noise, which can be a major issue to tackle on site or at a design stage. This paper show how standard Noise, Vibration, Harshness (NVH) post-processing can be used to better understand specific e-NVH (electromagnetic NVH) phenomena, and demonstrates some new post-processing to more quickly capture the root cause of e-NVH harmonics.

W. Siebenpfeiffer (Hrsg.): *Automotive Acoustics Conference 2019*, Proceedings, S. 226–237, 2020. https://doi.org/10.1007/978-3-658-27669-0_16

2 Standard NVH Post Processsing

2.1 Spectrograms

A spectrogram is a 2D representation of frequency versus rotating speed, displaying signal magnitude as a colormap (see Fig. 4). Spectrograms are generally obtained using a Short Time Fourier Transform of the time waveform of measured quantity, therefore requiring a tradeoff between frequency resolution and time resolution. They can be drawn on vibration levels (displacement, velocity or acceleration in radial, axial or tangential directions) or sound levels (in this case, the term sonogram or sonagram is sometimes used). They can also be applied to electrical machine phase currents.

Spectrograms are sometimes referred as waterfalls, Campbell diagrams or colormaps. The term waterfall should be limited to a 3D plot of a 2D spectrogram (taking the amplitude of the signal as z-axis to draw a 3D surface). The term Campbell diagram should be limited to schematic plots of excitation frequencies and structural natural frequencies without considering the magnitude of the system response. The abusive term "colormap" refers to the graphical post-processing.

All the excitations proportional to the speed cross the origin of the graph. If the run-up is linear (constant acceleration), these excitations appear as straight lines. This is the case of electromagnetic excitations due to pole/slot interactions, but not necessarily the case of electromagnetic excitations due to switching (Pulse Width Modulation), which can appear as V-shape excitations.

Structural mode natural frequencies appear as vertical bands in a spectrogram: even if electromagnetic excitations are purely harmonic, there is always a broadband excitation source coming from mechanical or aerodynamic forces which amplifies the response of the electric powertrain independently of speed. This vertical bands are therefore very informative; depending on the position of the sensor and the direction of the vibration, one may guess the modal shape that is associated to this particular frequency.

2.2 OT

OT technique consists in processing a spectrogram to extract signal magnitude at multiples of rotational frequency (named orders k). Both spectrograms and OT provide resonances and excitation electromagnetic forces. The equivalent of a spectrogram can be drawn as an ordogram, where the y-axis directly represents the orders k instead of the frequency.

Order tracking analysis allows to identify the respective frequencies of the electromagnetic excitations as $f = kf_R$ (using f: frequency of the excitation force, k: "mechanical order", f_R: mechanical rotation speed frequency). For instance, electromagnetic excitations may only appear at $48f_R$ and $96f_R$ (also respectively called H48 and H96) on a specific machine. This information is very valuable because there is a correlation between the frequency of the electromagnetic excitation and its spatial distribution along the airgap of the electrical machine – this spatial frequency is called a "wavenumber" (or spatial order) and named r, some example of wavenumbers are given in

Fig. 1. $r = 0$ corresponds to a pulsing vibration wave; $r = 1$ to an Unbalanced Magnetic Pull (UMP), the magnetic equivalence of a mechanical imbalance (that would typically occur at H1 if the rotor is not mechanically balanced); other vibration patterns due to electromagnetic forces always rotate, even in case of Pulse Width Modulation (PWM).

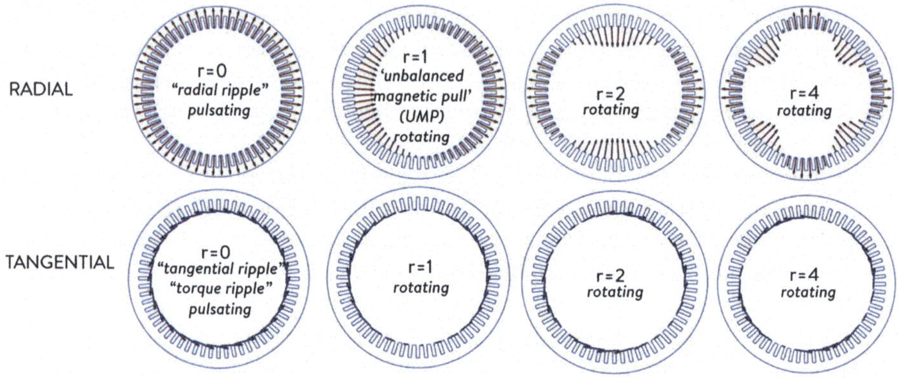

Fig. 1. Examples of different patterns of electromagnetically-excited radial vibration waves along the electrical machine circumference ($r = 0$: pulsating, $r = 1$: UMP).

The correlation between the excitation frequency f and the wavenumber r noted (f, r) can be known analytically but it depends on the load state, e-motor topology, airgap roundness and overall magnetic and geometrical asymmetries. Numerical simulation with MANATEE software can be also used to make this correlation.

Using the OT, the plot of the magnitude of a given vibration velocity order as a function of speed can give a clue on the presence of resonance. In particular, if the electrical machine is supplied so that magnetic forces remain constant over the full speed range (case of induction machine during constant flux operation, or permanent synchronous machine during open-circuit operation or fixed current angle and magnitude operation), the order tracking analysis of a radial accelerometer on the stator stack gives an image of the dynamic mechanical transfer function of the stator, and both natural frequencies and damping can be extracted (Fig. 2).

An important feature of e-NVH is that large resonances occur at two conditions: the electromagnetic excitation frequency must match a natural frequency, and the electro-magnetic excitation wavenumber must match the corresponding modal shape of the stator or rotor. This means that in a spectrogram, some excitation orders might cross some natural frequencies (vertical bands) without resonance. Therefore, the combination of the Order Tracking analysis and the identification of the natural frequencies present in the spectrogram can give some clues of the electromagnetic force wave-number responsible for a given resonance.

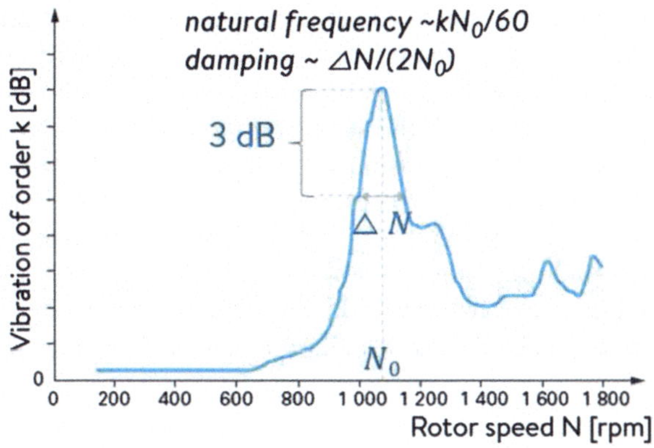

Fig. 2. Example of an Order Tracking analysis at H48 using a radial accelerometer on a stator.

2.3 ODS

An ODS consists in visualizing structural deflection shape on chosen degrees of freedom at a certain frequency and a certain speed, the structure being excited by its internal operational forces.

When an ODS is carried around the e-motor stator or rotor circumference at a frequency where driving forces are mainly of electromagnetic origin, one can visually analyze the nature of the vibration wave (rotating or pulsing) and the rotation direction. Associated vibration patterns correspond to either excitation force patterns (wavenumber r, far from resonance) or modal deflection shape (at resonance). An ODS can therefore confirm the main wavenumber of a radial or tangential vibration wave of the stator yoke. Note that the distribution pattern of vibration may not only involve a single wave-number, for instance due to eccentricities. Besides, the number of accelerometers to be put around the stator circumference must be chosen depending of the maximum wave-number to be captured according to Shannon theorem (e.g. 20 accelerometers to capture a wavenumber $r = 10$).

2.4 EMA

An EMA consists in externally exciting a structure while monitoring input force and structure acceleration. It identifies modal shapes, natural frequencies and modal damping of the studied structure.

The EMA allows to identify in "cold" conditions (no electromagnetic loading, no thermal effect) all the structural modes of the structure. The structural modes which are the most easily excited by electromagnetic forces are the radial deflection shapes of the stator yoke (for an outer stator topology) and the bending modes of the rotor shaft, as well as torsional modes of the stator and rotor. EMA results can be used when interpreting spectrograms, as the vertical bands can be associated to structural modes identified in the EMA. Generally, in EV/HEV electric traction applications, the lowest natural frequency is the ovalization mode (potential resonance with $r = 2$ magnetic forces) of the stator stack or coupled rotor/stator bending modes, and the highest frequency is the breathing mode of the lamination stack (potential resonance with $r = 0$ radial forces).

EMA naturally characterizes all the structural modes of the system, even the ones which may never be excited by operational magnetic forces. Besides, the operational conditions (boundary conditions, electromagnetic & inertial loads, temperature) may not be respected. Modal damping highly depends on temperature in electrical machines, so the interpretation of damping measured in cold condition might be limited. These limitations can be removed using Operational Modal Analysis (OMA) based on several run-ups to excite the structure on a wide frequency range.

2.5 Limitations

Those methods present some limitations to quickly and efficiently:

- discriminate wavenumbers r of the electromagnetic excitation waves and quantify their respective contribution,
- discriminate the physical origin of the electromagnetic excitation harmonics,
- identify the only structural modes which are excited by electromagnetic forces under operational conditions.

3 Advanced NVH Post Processing

3.1 Spatiogram Post-processing

A "spatiogram" is based on the spectrogram visualization of vibration waves filtered by wavenumbers [1]. For each wavenumber r, a discrete angular FFT is calculated from acceleration synchronized time signals (see (1)), then a standard STFT is applied to the resulting time signal.

$$A^r(t) = \frac{1}{N_{acc}} \sum_{i=1}^{N_{acc}} e^{j\alpha_i r} a_i(t) \tag{1}$$

where N_{acc}: Number of accelerometers, a_i: Angle of accelerometer i [rad], r: Circumferential wavenumber, $a_i(t)$: Radial acceleration of accelerometer i at time t [m/s^2].

Spatiogram allows to overcome limitations presented in Sect. 2.5. They can be realized both along radial and tangential directions and can be extended to the electrical machine length to capture longitudinal wavenumbers.

3.2 Flux Sensor Post-processing

A search coil wounded around a stator tooth allows to capture the induced voltage by time-varying magnetic flux. This signal can be integrated in a standard NVH acquisition system (similarly as acceleration is integrated to obtain velocity) to get a picture of the average radial flux density over a given stator tooth. The radial magnetic force applied to stator teeth can be approximated as the square of the average radial magnetic flux.

A Frequency Response Function can therefore be estimated from the ratio between radial displacement and radial tooth force.

3.3 Current Weighting Post-processing

Maxwell stress electromagnetic forces are a quadratic function of the flux density. Depending on load state and electrical machines topology, excitation force might be proportional to the current or to the current squared. To obtain the structural response of the e-motor independently from variations of the excitation levels, one can therefore normalize the NVH levels to the current or the current squared to ease interpretations of resonances.

4 Application

4.1 Experimental Analysis

A PMSM machine is equipped with accelerometers, a microphone and a tachometer. No-load run-ups and ODS are carried. Several post-processings are done to identify vibration waves responsible for acoustic noise, namely spectrograms, order tracking analysis, operations deflection shapes and "spatiograms" (Fig. 3).

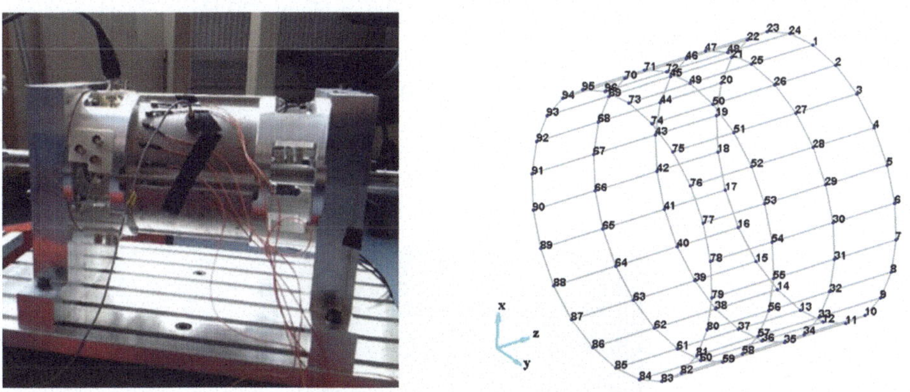

Fig. 3. Set-up with 6 accelerometers around yoke circumference (left), associated 24 × 4 meshing (right).

Spectrogram. Rotation speed of the PMSM is increased linearly between 0 and 1200 RPM over 40 s. Acceleration of the outer yoke in radial direction is recorded during the linear ramp-up. The resulting spectrogram is depicted in Fig. 4.

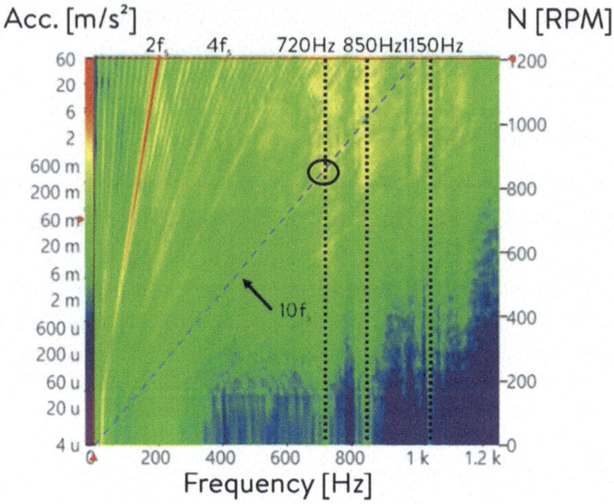

Fig. 4. Radial acceleration spectrogram.

The main forced excitation which appears as straight line crossing the origin of the graph, occurs at $2f_s$ (f_s, electric frequency). Other excitations occur at $4f_s$ and at $10f_s$. Three distinct structural mode natural frequencies of the stator yoke appear as vertical bands on the spectrogram: modes at 720 Hz, 850 Hz and 1150 Hz. A resonance between a force excitation occurring at $10f_s$ and the mode at 720 Hz is clearly noticeable at 850 RPM.

Ordogram. The equivalent of the spectrogram in Fig. 5 is pictured as an ordogram below.

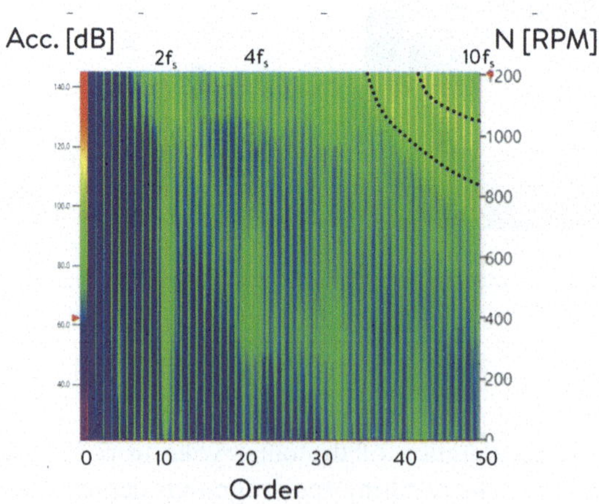

Fig. 5. Radial acceleration ordogram.

Same conclusions as for the spectrogram can be drawn from the ordogram.

Order Tracking Analysis. An Order Tracking at the order 50 is performed to investigate further the resonance at $10f_s = 10pf_R = 50f_R$ (f_R: mechanical frequency: p: number of pole pairs). Both natural frequencies and damping factor can be extracted (Fig. 6).

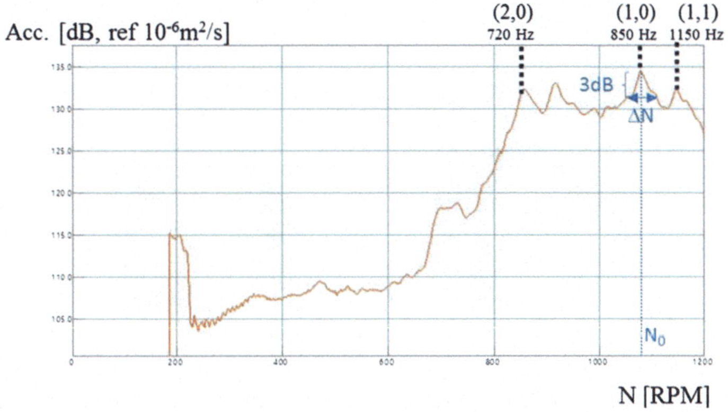

Fig. 6. Order Tracking Analysis of order 50 from acceleration spectrogram.

ODS. Vibrations are measured using 6 roving accelerometers and 1 reference accelerometer located on the outer yoke of the stator. The 6 roving accelerometers are moved 16 times to map the 96 nodes of the whole geometry (from Fig. 3 (right)). With this setup, the ODS can capture radial modes with circumferential wavenumbers up to 12 and axial wavenumbers up to 2 according to Shannon theorem (24 nodes per circumference plane and 4 nodes per axial plane). The forced response of the SPMSM stator is investigated at 550 RPM where no resonance is noticed on the acceleration spectrogram. Deflections due to the forced stress harmonics $2f_s$, $4f_s$, $10f_s$ are illustrated in Fig. 7. There are rotating waves of respective wavenumber −2, −4 and 2. The rotation direction is captured with the wavenumber sign (positive or negative). ($2f_s$, −2) and ($4f_s$, −4) rotates in the opposite direction compared to ($10f_s$, 2). The ODS accounts for another excitation at ($2f_s + f_R$, −1) which typically highlights a dynamic eccentricity.

Spatiogram. The testbench is equipped with 20 accelerometers even evenly distributed along the yoke circumference to study the contribution of each wavenumber up to a maximum wavenumber $r = 10$ according to Shannon theorem. A run-up is performed, and averaged spectrogram is filtered by wavenumber from −9 to +10. Spatiograms at −2, 2, −4, 4,−1 and 10 are given in Fig. 8.

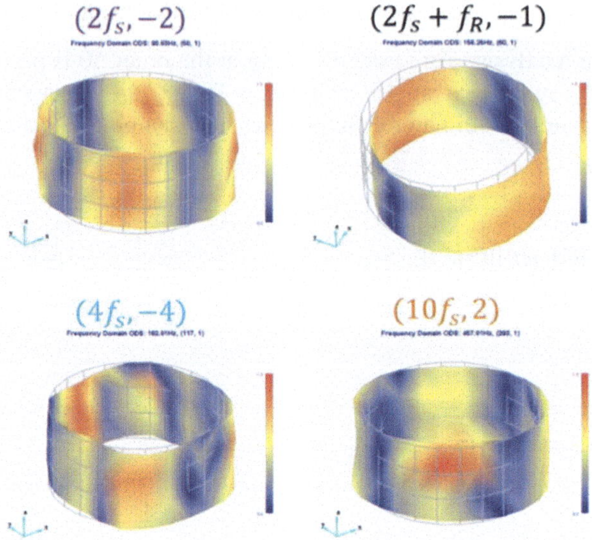

Fig. 7. Rotating deflections of the stator yoke at 850 RPM due to forced excitations at ($2f_s$, −2), ($4f_s$, −4), ($10f_s$, 2) and ($2f_s + f_R$, − 1).

The largest contributor to a given excitation can be identified comparing all spatiograms. Significantly higher acceleration levels are noticed on the straight line at $2f_s$ for $r = -2$, meaning excitation at $2f_s$ actually mostly comes from $r = -2$. Similarly higher accelerations levels are noticed on the straight lines at $2f_s + f_R$ for $r = -1$ and at $4f_s$ for $r = -4$ meaning excitation at $2f_s + f_R$ mostly comes from $r = -1$, and excitation at $4f_s$ mostly comes from $r = -4$. No specific excitation occurs at wave-number $r = 10$.

The rotation direction is captured with the wavenumber sign (positive or negative). In addition, structural modes that are excited by electromagnetic forces can also be captured on a spatiogram. Modes of deflection shape (1,0) (and respectively (2,0)) actually appears as straight lines on spatiograms at $r = 1, -1$ (and respectively $r = 2, -2$).

4.2 Simulation

Electromagnetic and vibroacoustic numerical calculations are run within MANATEE simulation environment at variable speed and at specific speeds [2]. Magnetic excitation forces are given as output from the simulation and are expressed as stress harmonics applied at the interface between the airgap and the external stator structure.

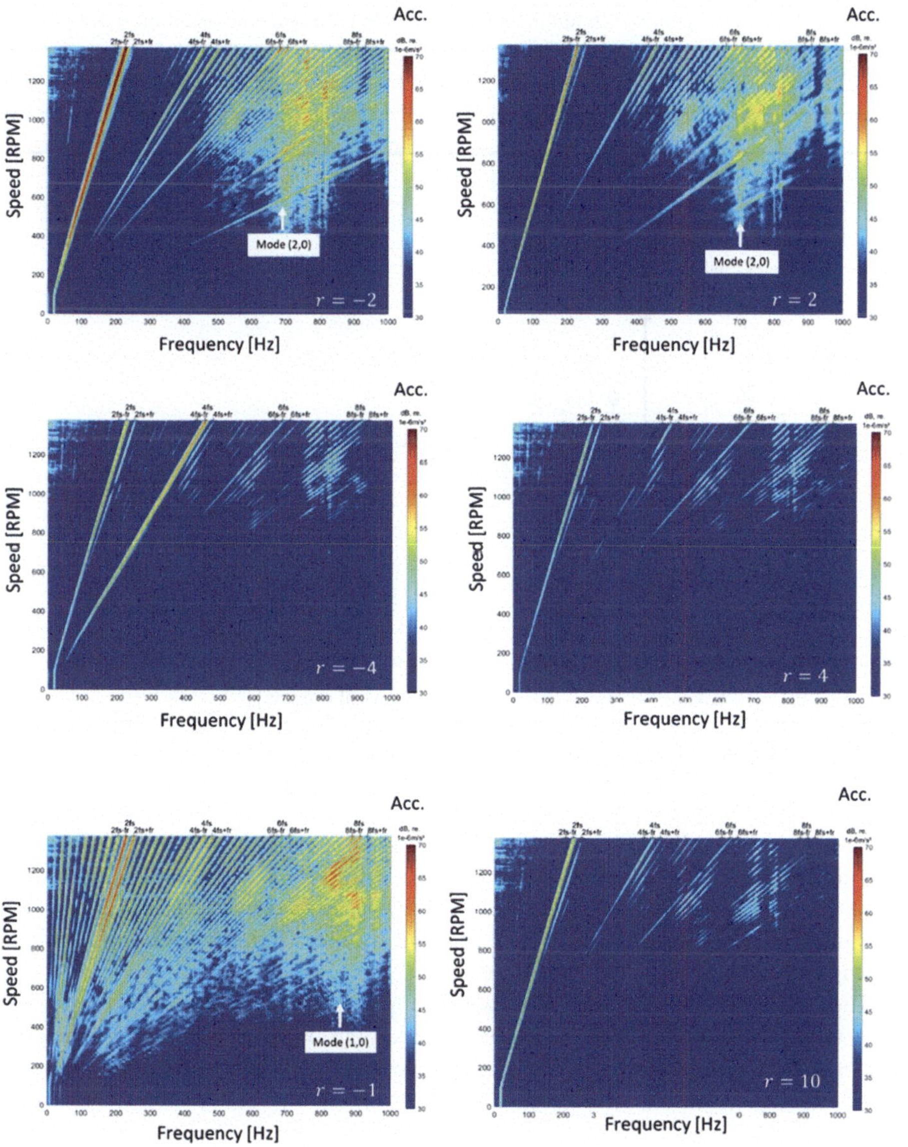

Fig. 8. Spatiogram at wavenumber −2, 2, −4, 4, −1 and 10

Main stress harmonics calculated for the testbench PMSM are represented in Fig. 9 as a function of electrical order k and wavenumber r (Table 1).

An additional stress harmonic appears at $(2f_s + f_R, -1)$ when dynamic excentricity is considered.

Simulation results are favorably compared to experimental results for $(2f_s, -2)$, $(4f_s, -4)$, $(10f_s, 2)$, and $(2f_s + f_R, -1)$.

Table 1. Main air gap stress harmonics of testbench PMSM machine and their main origins.

Stress harmonics σ_i	σ_1	σ_2	σ_3	σ_4
(kf_s, r)	$(2f_s, 10)$	$(2f_s, -2)$	$(4f_s, -4)$	$(10f_s, 2)$

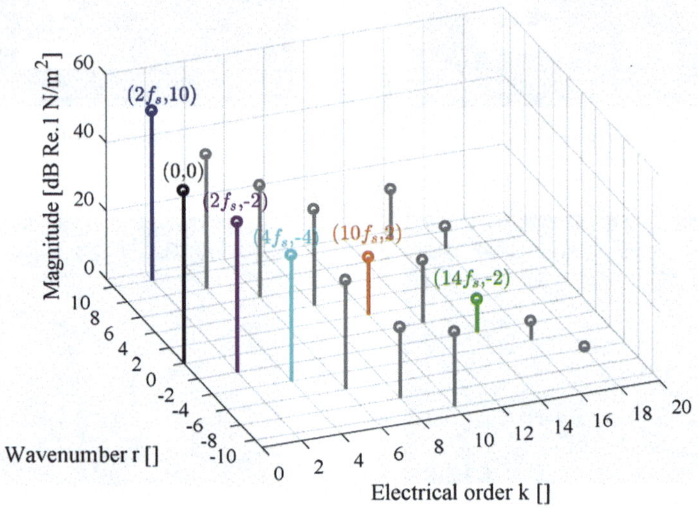

Fig. 9. Main air gap stress harmonics of testbench PMSM machine

However, the deflection due to the fundamental stress ($2f_s$, 10) is not experimentally noticeable at $2f_s$. This is explained by the stiffness of the stator structure which filters the main stress harmonic of wavenumber $r = 10$, and by the fact that this stress harmonic of wavenumber $r = 10$ is modulated by the 12 stator teeth due to spatial aliasing and folded back to wavenumber $r = -2$.

These magnetic stress harmonics which applies on the stator structure are computed using the Maxwell Stress Tensor in the airgap, which expresses the quadratic relationship between magnetic stress and magnetic flux density distribution, such as:

$$\sigma_r = \frac{B_r^2 - B_\theta^2}{\mu_0} \tag{2}$$

$$\sigma_\theta = \frac{B_r B_\theta}{\mu_0} \tag{3}$$

Using the convolution approach [3], each stress harmonic can be decomposed into a sum of combination of flux density harmonics, which enables to understand their origin and troubleshoot magnetic noise issues in electrical machines. For example, the origin of stress harmonic ($10f_s$, 2) is illustrated in Fig. 10, which shows the five first combinations of flux density harmonics contributing to this specific stress harmonic.

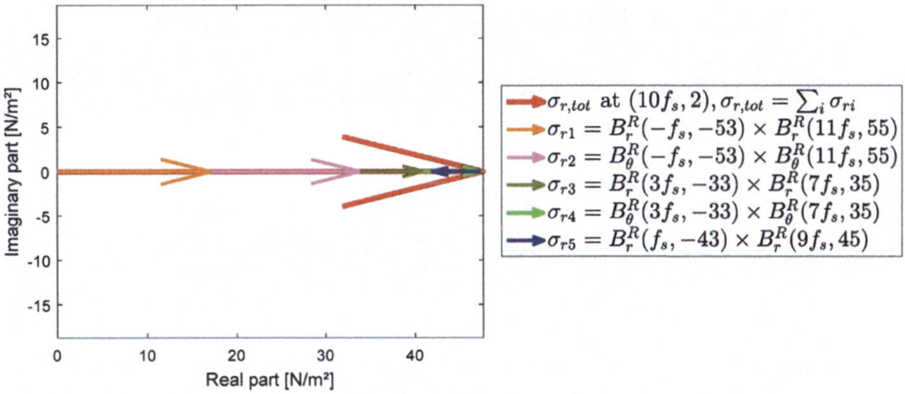

Fig. 10. Decomposition of radial stress wave harmonics ($10f_s$, 2) into combinations of flux density harmonics

5 Conclusion

Some advanced NVH post processing techniques are proposed to fully analyse the origin of electromagnetic noise and vibrations present in electrified powertrains. Unlike standard NVH techniques, they allow to efficiently discriminate the wavenumbers (spatial orders) of the magnetic excitation waves and to quantify their respective contribution, speeding up the diagnosis and the choice of noise mitigation strategies to be implemented. Other post processing based on tooth flux sensor can be used to improve numerical simulation models and quantify operational asymmetries of the e-machine.

References

1. Q. Souron, J. Le Besnesrais, M. Hecquet: Analysis of electromagnetically-induced vibrations of electrical machines based on spatiogram technique, ISEF 2015.
2. MANATEE: Magnetic Acoustic Noise Analysis Tool for Electrical Engineering, version 1.08.1 (Feb2019 Build), http://www.eomys.com, EOMYS ENGINEERING, 2019.
3. R. Rothe, M. van der Giet, and K. Hameyer, "Convolution approach for analysis of magnetic forces in electrical machines," *COMPEL – Int. J. Comput. Math. Electr. Electron. Eng.*, vol. 29, no. 6, pp. 1542–1551, Nov. 2010.

Auxiliary Equipment in Electric Vehicles

Evaluation of Noise Quality Based on Component Tests

Sebastian Roßberg(✉)

HEAD acoustics, Ebertstr. 30a, 52134 Herzogenrath, Germany
sebastian.rossberg@head-acoustics.de

Abstract. This paper presents tools and methods that can be used to track the noise quality of auxiliary equipment in the development process without extensive vehicle measurements. Its aim is to provide a guideline for the evaluation of sound quality on component test benches and to sensitize engineers to the challenges of the individual steps. Besides the identification of relevant noise components and the results of test bench evaluations, the focus of this paper is on the determination of characteristic values. The selection of specially tuned psychoacoustic analyses and a suitable combination of characteristic values are demonstrated using examples of tonal noise components.

Keywords: Noise quality · Test bench · Auxiliary equipment

1 Introduction

Suppliers of auxiliary systems are increasingly confronted with noise quality requirements placed on their products. During the development process, it is only rarely possible to analyze the acoustic quality obtained in as-installed state in the vehicle; evaluation of the isolated system on a corresponding component test bench is the preferred approach, due to its considerably lower cost.

However, as suppliers have very often a limited NVH experience, it is important, on the one hand, to make them aware of the complex task of noise evaluation and, on the other hand, to support them in establishing the corresponding know-how and in developing efficient solutions. Following is a description of which procedure ensures reliable and process-assured test bench results to allow predictions about sound quality in the later built-up vehicle to be made.

1.1 Challenges and Assessments

In terms of sound quality of auxiliary systems, the reference should always be the auditive assessment of the operating noise in the vehicle by the end user. Although this is a highly complex situation where many factors influence the assessment, it is the only way to capture the true relevance of individual noise components.

W. Siebenpfeiffer (Hrsg.): *Automotive Acoustics Conference 2019*, Proceedings, S. 238–245, 2020. https://doi.org/10.1007/978-3-658-27669-0_17

Therefore, it would seem likely to measure and evaluate the noise of components in-situ in the vehicle only. However, due to the time and effort for such measurements, it is not reasonable to perform these for a larger number of samples or on a regular basis. Nevertheless, they are necessary to get required reference signals or to control noise quality by random sampling.

The construction of test benches that create conditions identical to the in-situ situation is in most cases also not realistic, since all influencing parameters of the transfer paths in the vehicle have to be reproduced. Especially the inclusion of the airborne paths results in requirements to the measurement surroundings that cannot be satisfied in the workshop or production locations.

The most feasible way is therefore to use test benches with a sensing as simple and robust as possible but as well capable to detect all relevant phenomena that can lead to noise problems in the vehicle. Based on this scenario, the goal of good acoustic quality in the vehicle can be transferred to isolated components on the test bench with the application of various strategies. The following approach has proved its value in numerous projects (see Fig. 1):

- identification of the relevant noise components
- feasibility study, consisting of test bench evaluation and characteristic value determination
- testing and implementation on the test bench.

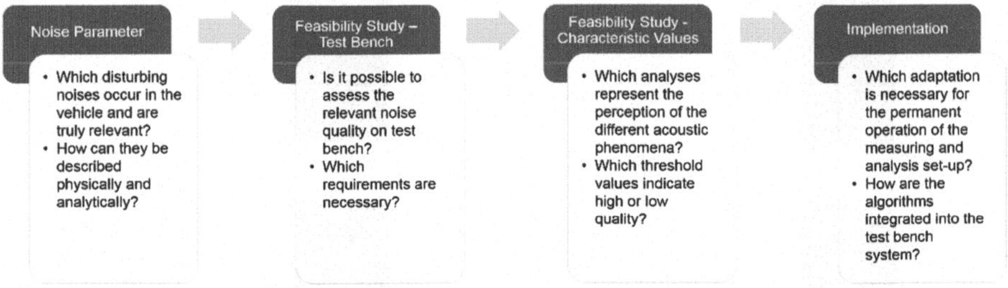

Fig. 1. Procedure for development of a test bench capable of assessing noise quality in the vehicle based on measurements of the isolated component

2 Identification of Relevant Noise Components

The objective of the initial step is to identify both critical operating situations and the noise contributions made by them as accurately as possible based on the reference situation in the vehicle. This provides the precondition for efficient and relevant processing of the next steps. The compilation of this list should be done with great care to capture all relevant noise components. The characteristic values to be developed depend on this list; the detection of further noises cannot be guaranteed. Moreover, a supplier can use this catalogue to discuss his approach to sound quality testing with his customer and gain mutual agreement about its extent.

Highly comprehensive methods exist to identify which noises are relevant [1]; however the majority of suppliers is not in a position to apply them. Suppliers often face more or less special requirements from their customers. Therefore, they may already be aware of conditions or noise phenomena that are conspicuous in the vehicle.

Measurement and analysis of interior noise complaints are the basis for the creation of a catalog of the various disturbing noise patterns in the vehicle. Each pattern receives a unique name to prevent any confusion, and is characterized by its typical acoustic properties in the time and frequency domain. By this, they can be compared with the test bench measurements. Table 1 shows an exemplified noise catalog with the four noise patterns clack, squawk, chirp, and grind.

Table 1. Typical example of a noise catalog: the auxiliary system in question is characterized by four different disturbing noise patterns in the vehicle

Name	Diagram	Description	Name	Diagram	Description
Clack	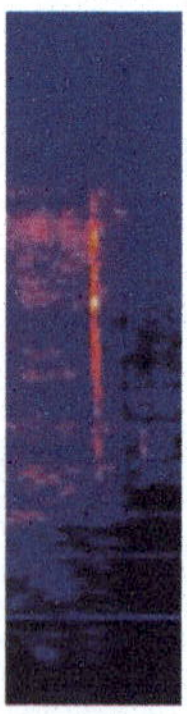	– Impulsive – 1–4 kHz – 1.5–3.3 Hz maximum level – Can occur periodically	Chirp	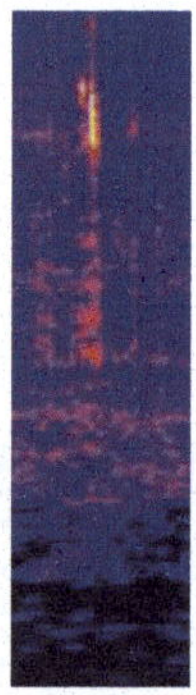	– Impulsive or <0.5 s duration – 4.5–7 kHz maximum level, parts above 10 kHz possible – Occurs only isolated
Squawk	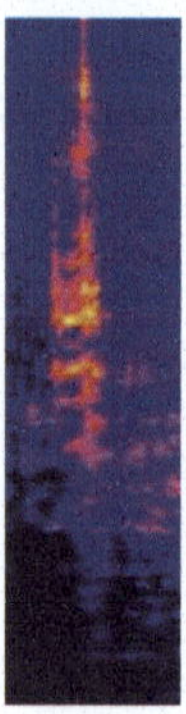	– 0.5–1 s duration – Tonal components with frequency shifts – 1.4–1.8 kHz maximum level, sometimes higher – Typical at the beginning	Grind	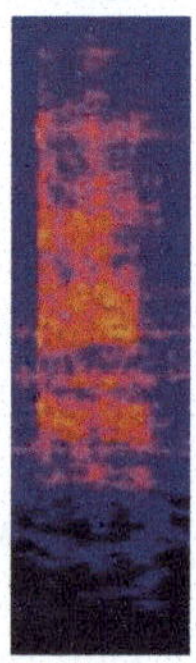	– 0.5–1.5 s duration – Broadband, noise-like – 0.6–2.9 kHz maximum level – Often periodically

3 Feasibility Study

As stated above, the assessment of the perceived airborne noise in the vehicle is always our reference. Although it is possible to measure and analyze airborne noise of isolated components, it involves significant overhead, as special isolated measuring cells are required. Hence commonly used component test benches

- are unable to recreate the transmission paths and emission characteristics in the vehicle,
- use less interference-prone variables for analysis; these are often component vibrations, which are recorded using accelerometers or using a non-contact laser Doppler vibration measurement device, or other operating variables, for example control voltages or currents.

3.1 Test Bench Evaluation

In this step, the objective is therefore to clarify the conditions under which the relevant acoustic phenomena, that give rise to interior noise complaints, can also be detected on a component test bench. In order to achieve a good correlation, it is advisable to produce mountings and operating states on the test bench that are comparable to those in the vehicle, and to select measured variables that causally relate to the noise emission.

Performing these considerations as early as during the test bench design phase provides the advantage that the test bench requirements (in terms of sensor technology, environment, etc.) can be defined relatively freely. Under certain circumstances, however, there is limited experience of the noise quality existing in this phase; or not all complaints or customer requirements are yet known.

If, in contrast, a test bench is already in operation and producing only unsatisfactory test results, clear expectations generally exist in terms of the detection capabilities of an optimized test bench. The objective is an improved – or additional – assessment of specific acoustic phenomena. During test bench evaluation, it must first be established whether this goal can be achieved at all with the existing setup. If this is not the case, additional requirements must be defined. The NVH engineer must carefully review the relevance of these requirements to the assurance of product quality and then clearly present the requirements; after all, significant changes to a test bench already in operation are rarely welcome and often difficult to implement.

3.2 Determination of Characteristic Values

Development of a reliable characteristic classification or metric requires a comprehensive data basis, on which all subsequent analyses and evaluations are based. Only the data contained can be adequately considered in subsequent analytical evaluation. Missing disturbing noises cannot be included.

To create the data basis, all acoustic phenomena must be recorded in their different forms – from the unacceptable to the barely perceptible. An adequate number of noises about which no complaints have been received should also be available as a reference. The respective components must be measured both in the vehicle and on the test bench and all relevant signals, including required control parameters, taken into account. This allows the subsequent establishment of the correlation between different measured variables.

Evaluations. Auditive evaluation of vehicle measurements is used for the subsequent adaptation of characteristic value parameters and to define limits. Where no other specifications exist, these evaluations can be determined using carefully prepared listening tests [2]. In practice, suppliers often focus on specific scenarios, for example:

- evaluation by experts in parallel with vehicle measurements
- customer requirement specifications
- joint identification of criteria with the customer
- okay and not-okay evaluations by production employees.

In cases where an already established procedure of auditive evaluation of component noise is to be extended or replaced by analyses, the reference signals can derive deviational not from a vehicle but the existing test bench. All methods have in common that they supply an additional parameter to every reference measurement – a classification regarding its sound quality.

Selection of Suitable Analyses. In addition to the evaluations, the various manifestations of disturbing noises in test bench measurements must be quantified. However, simple global variables, such as sound pressure level, are not sufficient for doing this. Generally, component noise is more subtly perceived and its quality assessed using characteristics such as conspicuousness or value perception. Reference must also be made to adequate psychoacoustic and specially developed algorithms to quantify these perceptions.

The selection of suitable analyses and algorithms to evaluate test bench measurements always depends on the noise phenomena complained of in each case. In order to correctly capture the respective characteristic, it is advisable to answer the following questions for each individual phenomenon:

- What is conspicuous about this contribution? Using the evaluations collected and one's own auditive analysis, an acoustic description of the noise contribution is created. Indications for this can be found in the noise catalog.
- How can these characteristics be modeled? The descriptions created before must then be analytically implemented into filters, analyses, and specific calculations. To do this, the experience and knowledge of the NVH engineer in the field of signal processing and psychoacoustics is required.

Figure 2 shows an example for the selection of suitable analyses. The objected noise component is a gear whining – auditive prominent especially in the range of 4–10 s and 18–20 s. The aim was to find an analysis that could identify the whining without the use of any further information (like engine speed or frequency range). Based on evaluations and in-house expertise, the acoustically conspicuous (tonal) orders are responsible for the phenomenon. Various analyses of the tonality were compared. The specific tonality according to DIN46681 could only detected the tonality between 4–10 s and was thus excluded. The specific prominence ratio clearly detected all relevant tonalities but evaluated also areas as tonal that are auditive inconspicuous and was excluded too. The two remaining analyses looked booth promising in der 3D diagram. But for the application of limits, a 2D diagram or a single value is needed. Here, only the specific tonality (hearing model) could create an appropriate diagram. In conclusion the best match was achieved with the specific tonality (hearing model); the 2-D representation of which can be used for comparison with the evaluations to identify the characteristic value.

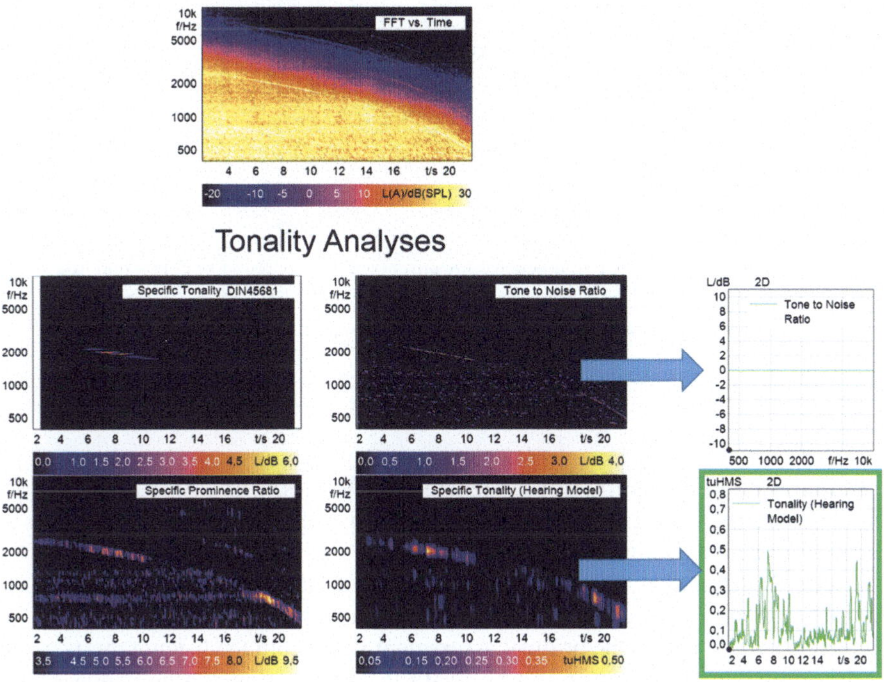

Fig. 2. Selection of suitable analyses – As an example displayed for the conspicuity "gear whining"

When selecting the analytical descriptions, it can be said that the more specific the calculation, the more accurate the quantification of the corresponding noise contribution. This means, however, that the calculation is more susceptible to deviations of noise properties or component changes. In contrast, more general global measures are more resistant to changes; but can only describe non-specific manifestations. Under standard conditions, a specific characteristic value will provide more information; however with slight parameter changes, errors may go undetected. In the example in Table 2, the frequency pattern is indicative of the defect of a specific part A. Under standard conditions, the specific characteristic value provides additional information about the defective part. However, if the test parameters change – as, in this case, the speed as a result of increased friction – the error can only be detected by the global characteristic value – without part allocation.

Table 2. Under standard conditions, a specific characteristic value provides more information; however with slight parameter changes, errors may not be detected – the global characteristic value allows no part allocation

Limit	Damage at Part A standard		Damage at Part A Parameter change	
Specific	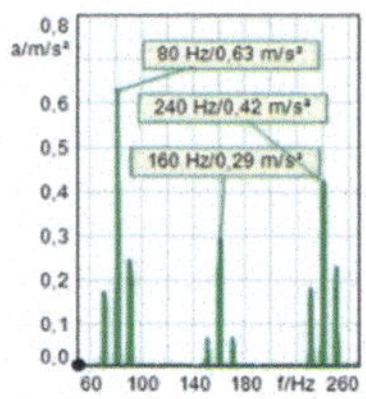	Limit 80/160/240 Hz: 0,3 m/s^2 80 Hz: NOK 160 Hz: OK 240 Hz: NOK Report Damage at Part A	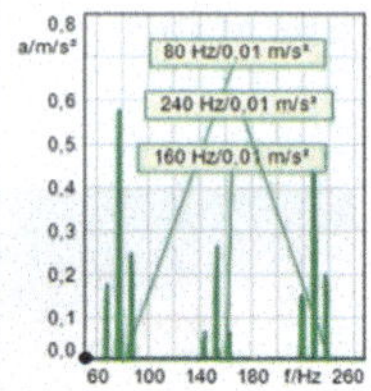	Limit 80/160/240 Hz: 0,3 m/s^2 80 Hz: OK 160 Hz: OK 240 Hz: OK Report Part A OK
Global	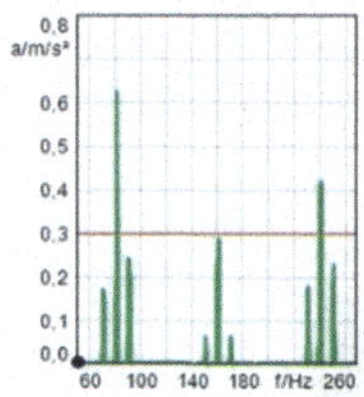	Overall limit: 0,3 m/s^2 Overall: NOK Report Damage without allocation	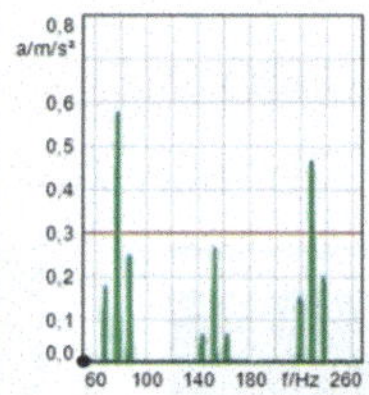	Overall limit: 0,3 m/s^2 Overall: NOK Report Damage without allocation

Combination and Evaluation of Characteristic Values. For a final result, the specific characteristics must be combined to form an overall analytical evaluation that describes the sound quality of each component.

In serial combination, the individual characteristic values are combined in the form of a metric using linear combination to form a global characteristic value [2]. This will then generally reflect the original assessment scale from the listening tests, so that the corresponding classification can be read off directly. How the individual characteristic values contributed to the overall result cannot be read off directly due to the different value ranges; however, it can be integrated via additional considerations.

For parallel combination, in this case, each single characteristic value is evaluated individually and the overall assessment derived. As a worst-case criterion, it can be equivalent to the worst individual result. This method is typically used when direct limits can be applied and the origin of the overall assessment is of interest. Due to the higher information content, the overhead is also greater than in the serial method. For each noise phenomenon, the respective auditory assessments are required to derive individual scaling or limits.

3.3 Test and Implementation

In the final step, the intent is to set up the test and the characteristic value calculation on the test bench for permanent use. The degree of automation here depends on the individual conditions of the test bench as well as on the users. First, a fixed procedure must be defined from the application of the test object on the test bench to the test measurement and the output of the results. This is then executed on a large number of test objects, where once again it is ensured that the entire spectrum of potential disturbing noises is detected. This proof of concept phase is used to refine the test bench procedure for stable operation and to verify the metrics or characteristic values.

4 Conclusion

If the sound quality of auxiliary systems has to be evaluated analytically on component test benches, some significant challenges exist; these however can be overcome with an appropriate methodology and NVH experience. This ensures the creation of a reliable analysis tool for the continuous evaluation of acoustic quality within the development process on component test benches.

References

1. Sellerbeck, P.: Eine innovative Methode zur Fahrzeugbewertung und Zielgeräuschdefinition. In: 8. Magdeburger Symposium Motor- und Aggregateakustik, pp. 54–72, Expert-Verlag, Renningen-Malmsheim (2014)
2. Fiebig, A., Genuit, K.: Hörversuche und Metrikentwicklung. In: Genuit, K. (ed.): Sound-Engineering im Automobilbereich. Springer, Berlin/Heidelberg (2010)

Zusatzmaterial

Wolfgang Siebenpfeiffer(✉)

Stuttgart, Deutschland
wolfgang@siebenpfeiffer.eu

Zusammenfassung. Dieses Kapitel enthält zusätzliche Folien, die auf der Konferenz präsentiert wurden. Sie sind als elektronisches Zusatzmaterial auf SpringerLink frei verfügbar.

Elektronisches Zusatzmaterial Die elektronische Version dieses Kapitels enthält Zusatzmaterial, das berechtigten Benutzern zur Verfügung steht. https://doi.org/10.1007/978-3-658-27669-0_18

W. Siebenpfeiffer (Hrsg.): *Automotive Acoustics Conference 2019,* Proceedings, S. 246–246, 2020. https://doi.org/10.1007/978-3-658-27669-0_18

Tagungsbericht

Frank Jung

5. Internationale ATZ-Fachtagung Fahrzeugakustik

Die zunehmende Vielfalt und Komplexität moderner Antriebsstränge geht einher mit einem erhöhten Aufwand für die akustische Entwicklung. Wie sich diesen Herausforderungen begegnen lässt, wurde am 9. und 10. Juli 2019 in Rüschlikon bei Zürich (Schweiz) auf der Automotive Acoustics Conference diskutiert. 220 Teilnehmer aus 21 Nationen erlebten ebenso spannende wie interessante Vorträge.

Wichtige plattform für NVH
Alternative Antriebe, Leichtbaustrukturen und kleinere Motoren stellen Akustik-Ingenieure vor große Herausforderungen. Um der rasanten Dynamik des technischen Fortschritts gerecht zu werden, stehen die Entwickler unter anderem vor der Aufgabe, nicht nur die Geräusche des Antriebsstrangs, sondern auch vieler anderer Komponenten effektiv zu reduzieren. Auf der anderen Seite gilt es an manchen Stellen aber sogar, wegfallende Geräusche zu substituieren. Leichtbaustrukturen, alternative Antriebe und kleinere Motoren bringen dabei also in beiderlei Hinsicht neue Hürden auf dem Weg zu einem komfortablen NVH-Level mit sich. Bessere Messanalysen und schnellere Berechnungsmethoden sind dabei wichtige Werkzeuge und zentrale Themen in der Expertendiskussion.

Die vielleicht wichtigste Plattform für diese Diskussion stellt die gemeinsam von ATZ und Autoneum veranstaltete Automotive Acoustics Conference dar, die mit 220 Teilnehmern in Rüschlikon stattfand. Die Hälfte der Fachbesucher kam aus nicht-deutschsprachigen Ländern, was den internationalen Charakter der Konferenz unter streiche, wie Dr. Alexander Heintzel, Chefredakteur der ATZ-MTZ-Gruppe, in seiner Begrüßung hervorhob. Die längsten Anreisen hatten die vielen Teilnehmer aus Japan – die auch die größte ausländische Gruppe stellten – sowie aus China, Taiwan, Südkorea, Indien und Südafrika.

Neue störende Geräusche
Dr. Léon Gavric, Vice President, Senior Expert NVH der PSA Groupe, prognostizierte als erster Redner, dass der zunehmende regulatorische Druck auf Verbrennungsmotoren, kombiniert mit Verbesserungen bei elektrischen Antrieben und Batterieleistungen, zu einer größeren Nachfrage bei Elektrofahrzeugen führen werde. Das Fehlen einer Maskierung vieler Geräusche durch den

W. Siebenpfeiffer (Hrsg.): *Automotive Acoustics Conference 2019*, Proceedings, S. 247–249, 2020. https://doi.org/10.1007/978-3-658-27669-0

Verbrennungsmotor werde diverse Systeme und Geräte, Windgeräusche, Rollgeräusche und Reifenrauschen stärker in die Wahrnehmung rücken. Entsprechend groß seien die NVH-Anforderungen im Vergleich zu Fahrzeugen mit Verbrennungsmotor. Eine neue, aber entscheidende NVH-Herausforderung bei Elektroautos sei es, das „Jammern" im Elektroantrieb zu kontrollieren. Störende Geräusche wie Quietschen, Rasseln und Summen seien im leisen BEV-Bereich ebenfalls verboten.

In die gleiche Kerbe hieb Dr. Ralf Bocksch, NVH Project Coordinator, Daimler, und verantwortlich für die Elektromarke EQC von Mercedes-Benz: „Der Sound macht den Unterschied". Die größte Aufgabe sei die Reduzierung des Straßenlärms: Obwohl die Batteriemasse positive Auswirkungen auf die Karosseriestruktur hat, führen härtere Fahrgestelle und Reifen zu mehr akustischen Auswirkungen von der Straße. Erfahrungen aus aktuellen Elektroautos von Mercedes seien der Ausgangspunkt für die NVH-Entwicklung des neuen EQC gewesen. Entsprechend dieser Basis wurden Dämpfungs- und Isolierteile entwickelt, indem das strukturelle Konzept vom Modell GLC übernommen wurde. Nach einem stringenten, zielorientierten Entwicklungsprozess sei es mit dem EQC gelungen, einen leisen, komfortablen und charakteristischen Klang zu entwickeln.

Leise Lagerung

Bei seinem Vortrag zu den Konstruktionsprinzipien der Lagerung im elektrischen Antriebsstrang ging Dr. Gregory Chen, Senior Chief Engineer, Geely Automobile Research Institute, bereits sehr in Details. Seinen Ausführungen zufolge seien die Resonanzeigenschaften der Montagehalterung wichtige Faktoren für das Geräusch des elektrischen Antriebs im Fahrzeug bei niedriger Geschwindigkeit (circa 40 km/h oder weniger). Die Montagehalterung müsse kompakt gebaut sein, und je höher die Modalfrequenz, desto besser isoliere das Befestigungssystem die Vibrationen. Chen empfahl hier eine Frequenz von größer als 1000 Hz sowie die Vermeidung einer Kopplung der Modalfrequenz der Montagehalterung und der internen Resonanzfrequenz des Montagegummis.

Simulation und Künstlicher Sound

„Verfahren und Lösungen für die Integration der HV-Batterie in die NVH-Entwicklung eines BEVs", lautete der Titel des Vortrags von Dr. Francesca Ronzio, Team Leader Products and Systems Simulation, Autoneum Management. Ihre beruhigende Nachricht für viele Entwickler lag darin, dass bestehende und etablierte CAE-Systeme Methoden, wie zum Beispiel die Statistische Energieanalyse (SEA) und Finite-Elemente-Methoden (FEM), das NVH-Design von BEV-Bodensystemen unterstützen können. Zu dem damit abgedeckten NVH-Paket gehörten nicht nur der Teppich und Dämpfungspads, sondern auch die Batterieabdichtung und Elemente zwischen Batterie und Karosserie.

Niemals ist jemand aus einem Elektroauto gestiegen und hat begeistert ausgerufen: ‚Wow, der Sound war ja fantastisch!', begann Ward recht kritisch. „Fragen Sie 15 Personen in einem Raum, wie ein Elektroauto klingen soll, und Sie bekommen

35 Antworten.“ Anhand realer Hörbeispiele beschrieb Ward den gesamten Entwicklungsprozess beim Fiat 500e von der On-Track-Bewertung der wichtigsten Merkmale eines „sportlichen Sounds“ bis hin zu den virtuellen Sounds.

Bedeutende Internationale NVH-Fachtagung
Die unter der wissenschaftlichen Leitung von Autoneum stehende und gemeinsam mit ATZlive organisierte Automotive Acoustics Conference gilt als eine der bedeutendsten internationalen NVH-Fachtagungen für Automobilhersteller, Zulieferer und Forschungsinstitute. Mehr als 20 Fachpräsentationen werden ergänzt durch Workshops zu neuen Mobilitätstrends. Eine umfangreiche Fachausstellung präsentiert aktuelle Innovationen aus dem Car-Akustikbereich.

Quelle: Jung, F. ATZ Automobiltech Z (2019) 121: 80. https://doi.org/10.1007/s35148-019-0116-x.

Autorenverzeichnis

W. Siebenpfeiffer (Hrsg.): *Automotive Acoustics Conference 2019,* Proceedings, S. 251–252, 2020. https://doi.org/10.1007/978-3-658-27669-0

The manufacturer's authorised representative in the EU is Springer Nature Customer Service Centre GmbH, Europaplatz 3, 69115 Heidelberg, Germany. If you have any concerns regarding our products, please contact ProductSafety@springernature.com

Printed and bound by CPI Group (UK) Ltd, Croydon, CR0 4YY
19/08/2026
02200177-0011